One Angel Left

PBY CATALINA "DUMBO"
Courtesy of the Stokes Collection, Carmel, California 1-800-359-4644

One Angel Left

By Jack Morris

Based on a True Story

Canyon Creek Books
Santa Rosa, California

One Angel Left

by Jack Morris

Canyon Creek Books
Santa Rosa, California

Copyright © 2001 by Jack Morris

First Edition

All rights reserved
No part of the contents of this book may be reproduced or transmitted in any form or by any means, electronic or mechanical, including photocopying, recording, or by information storage and retrieval system without written permission from the author. For information address Canyon Creek Books, 555 Montgomery Drive, C205, Santa Rosa, CA 95409

Library of Congress Catalog Card Number: 00-193006

ISBN 0-9706427-0-9

Special thanks to the Stokes Collection for permission to reproduce Stan Stokes' illustrations, PBY Catalina "Dumbo" on frontispiece and F4U Corsair on ending page of Epilogue.

Cover art by Jack Morris
Cover design and photo edits by Wendy Morris
Book design & production by Spring Hill Press, Sebastopol, California
Printed in Korea

PUBLISHED WORKS

One Angel Left
Personal Essays in AfterWords, an Anthology
and in the Los Angeles Times

For Rosalind and Nasira and those other Angelic Presences who created myriad miracles in my life in spite of my countless flaws, foibles, and frailties.

For Joe, Bob, Zeke, Leo, and Frank who remained in the realm of the Fallen Warrior south of the Equator, west of the International Date Line, in the Southwest Pacific.

*For **all** the faithful members of "The Flyin' Lion" Squadron, VMF 218, a Marine Fighter Squadron that flew F4U Corsairs in combat in World War II, all the ground crews, all our wives and loved ones.*

For the New Zealand Royal Airforce crews who spotted me and rescued me in Saint Georges Channel.

TABLE OF CONTENTS

PART I

CHAPTER 1	The Big Bang	1
CHAPTER 2	Working My Way Out	20
CHAPTER 3	The Getaway	24
CHAPTER 4	In Safe Hands	36
CHAPTER 5	Across the Pacific	57
CHAPTER 6	Pearl Harbor and Beyond	64
CHAPTER 7	Rough Seas Ahead	82
CHAPTER 8	The Big Hospital	88

PART II

CHAPTER 9	Burnham Home Capers	106
CHAPTER 10	An Ultimatum of Sorts	121
CHAPTER 11	Married at Last	126
CHAPTER 12	My Flight Status Threatened	131
CHAPTER 13	War Ended with a Big Bang	135
CHAPTER 14	Another Workout	139
CHAPTER 15	And Another	143
CHAPTER 16	Northward Bound	146
CHAPTER 17	Back to the Drawing Board	155
EPILOGUE	Ashes	164

ACKNOWLEDGMENTS

Thanks to the many who helped improve my writing style: classes and workshops conducted by Robert Kirsch and Miv Schaff from UCLA Extension, Nasira Alma of Oregon Writers Colony, and especially "Maudie" Sell, Arie Raysor, and Steve Boga, and the writers in the Anthology Group of Sebastopol, California, and to Laurette, who led me there. We all wrote this book. Thanks.

To officials and volunteer workers at the Kokopo Museum, East New Britain Provincial Government, Kokopo, Papua New Guinea, with special gratitude to Brian Bennett, I.D. Lindley, Alastair Norrie, and to Philip White of Blue Star Hirecars.

To Julia Cameron for writing "The Artist's Way" and making a huge difference in my whole life.

Thanks to Rosalind's late brother Raymond Vallens whose Trust helped finance the project.

To Rita Ter Sarkissoff of Spring Hill Press for final edit and make-ready.

To my late parents for furnishing great comfort, devotion, support, and enough gene power to "git it writ", thanks.

PART I

CHAPTER 1

THE BIG BANG

Four days slogged by since I was shot down on a large Japanese occupied island in the South Pacific with wounds that would need attention soon and signs of internal bleeding in my left eye. I'd put a large bandage on it and took sulfa to prevent infection in my torn left nostril and other lacerations, but didn't know if the drug was working.

I had no idea how long it would be before I reached a place where I could get the attention of Air-Sea Rescue. In earlier campaigns in the Solomons there were Australian coast-watchers and friendly natives who could snatch a downed flyer from under the noses of Japanese troops, radio for help, and arrange a pick-up point. Now, it was all up to me, and a lot more slogging. We had no coast-watchers here. In my survival kit was a set of instructions that told how to bribe a native, in Pidgin English, to help stranded flyers to a safe beach. But those natives of northern New Britain may have been too frightened of the Japanese, after two years of occupation, to be helpful or even trustworthy. I had no desire to meet them or to try my Pidgin English.

I was tired, felt soggy from the jungle's penetrating dampness. Sitting at the river's edge, thinking dark thoughts, weighed down with hopelessness; I had been trudging along this twisting, turning little river, wading in shallow places, walking along sandy beaches and gravel benches and hopping from boulder to boulder. Earlier I thought I saw smoke rising around the next bend and decided to bypass the bend by plunging into the almost impassable jungle and cutting across to the river beyond. In the gloom of the thick overhead canopy, and profuse undergrowth, I could neither see the vines tangling around my feet nor stinging nettles reaching out for bare skin. I stumbled over deadfalls that were laying at odd angles, tried to cut through the smaller stuff with the hunting knife my dad had made for me. Flying insects were buzzing around my face and hands. I finally stumbled through to the other side, vowing to never try that again. I searched upstream for the smoke and saw only rising vapor.

Peering back into the dark tangle where I had just been, I shuddered, brushed myself off, and started off again, downstream. As I rounded a sharp bend, wading in shallow, fast moving water, I skirted a low hanging limb, and found myself staring at the huge rear end of a wild pig rooting in a muddy backwash behind the tree. He didn't know I was there. Only five feet away, eating, he reminded me of my own hunger. Without thinking it through, I pulled out my .45 and fired my first shot in the combat zone directly at the innocent, bulky figure and missed! He bolted for the riverbank and disappeared in a flash.

Then I realized what a stupid thing I had done. Aiming to take the pig's life, something that would take hours of cooking, was incongruous. I had acted rashly because I was hungry. In four days, I had eaten only two small cans of pemmican—about twelve ounces. I needed fast food, the kind you can eat on the run. I had fired off a loud shot in a quiet jungle in enemy territory. *What was I thinking?*

My clothes were damp from an earlier rain and the wading. It was cloudy and gloomy and that foray into the thick tangle of jungle growth brought me down. The pig incident, my hunger, the apparent lack of progress and the stress of having to be ever watchful had me in a low mood. I was discouraged and distressed. *Will I get out of here alive?*

Then I heard her rich melodious voice with an answer. It was Roz. My wonderful one back home, thousands of miles away, calling to me. *Come on. You can make it.*

I looked up. I could see her face in the clouds. That dark hair, the light blue eyes, the gleaming smile. But above all, her look of concerned encouragement, of knowing all would be well. I felt relieved; with the burden lifted, energized, heartened by a loved one whose spirit had flown thousands of miles to aid in my escape from the enemy. Now, it was up to me, to get my legs moving.

As I walked along with renewed energy, head swiveling instinctively, eyes sweeping along both banks, my thoughts went back to this mystical encounter and I remembered our first meeting and how strange it had been.

On the Fourth of July, seven months ago, my friend Wanek and I were fresh out of flight school. We sported shiny gold wings and equally sparkling Second Lieutenant's bars, on our last fling before a hard squadron training session that lay ahead.

We were at the Ambassador Hotel in LA, the Officer's Club, a sort of USO for officers, a haven, an entertainment center, a place to get tickets and contacts for nights' lodgings.

Earlier that day, Wanek and I had attended a swim party arranged by the Officers Club at the Bel Air home of a celebrity. There were a lot of people; no one I knew. I never saw the star. It was a quiet Fourth, so far.

We were on our way back to our rooms when we decided to stop by the "O" club for no special reason, mostly out of habit. We walked down the long, lower esplanade, until we came to the main reception room. It was empty but the sounds of ping-pong drifted through the open door of the rec room. We sauntered into the room out of curiosity and there she was: this raven-haired beauty who turned and looked at me with incredible light blue eyes and a ready smile. There was a moment of recognition as I approached her. We were unmindful of the fact that we had just interrupted a game. Our eyes were locked. We knew each other. She handed her paddle to Wanek. He could finish the game with whoever waited on the other side of the table.

We were trying to figure out how we knew each other, so it was just a matter of getting acquainted by trying to remember where and when we had met. There was no clue. But did it matter? I kept looking into those light blue eyes. She had a definite Elizabeth Taylor, Millie Perkins look. That was probably it. I had seen her many times on the silver screen so I thought I knew her.

As a hostess at the club she had seen a lot of dark-haired Marine aviators, had even been to dances at some of the bases. I was not surprised she thought she knew me, but we had a nice reunion, at The Coconut Grove, later that evening.

The sun came out and hope returned. Back to boulder hopping and wading and trudging along the sparkling jungle stream. Shortly, I came upon a waterfall gently pouring into a pool by a warm, sandy beach. The conjunction of war zone and paradise seemed perfectly natural there. I took off my clothes and laid them out on the warm sand. The beach had a natural screen of bushes around it like the river willows back home. The water was refreshing, had a calming effect. I glided in under the waterfall. Strange how I could lie there, naked, yet not feel vulnerable, strange how I could be so close to danger yet have a feeling of profound peace. After this remarkable interlude, stretched out on the beach to dry, I thought back over the events of the four days gone by…

* * * * * * *

The Big Bang

On February 10, 1944 our formation of twenty planes of U.S. Marine Fighter Squadron 218 flew on a mission from Bougainville to Rabaul. We joined up with fighters from the U.S. Navy, U.S. Air Force and New Zealand Air Force to cover a flight of B25 Bombers in a bombing raid on Vunakanau, one of Rabaul's five airfields. To reach our target, we flew 220 miles over water and another twenty over a vast carpet of verdant jungle wilderness that comprised most of the Gazelle Peninsula on the Northern end of New Britain Island.

As we approached our target we could see some anti-aircraft fire, puffs of benign-looking black smoke off to our right, but nothing else of a hostile nature.

As we continued our approach, the fighter planes went into a weaving pattern at three levels. Our division flew at the middle level, in pairs, section leaders with wingmen together. The entire formation flew well past the target, then made a wide sweeping turn toward Rabaul, and things started to go wrong. The pair of Corsairs we were scissoring with took off in a straight line to cover the bombers. My section leader and I were in a tight right turn trying to catch up. At the moment I realized I was going to be tail-end-Charlie in this whole formation, I caught sight of a flight of about ten enemy fighters to our left and above. This meant that when we completed our turn they would be at six o'clock high, a very dangerous turn of events that called for a split-second decision. We could turn toward them and take them on or try to out run them or dive out and rejoin the formation farther on. The split-second was all we had and before I could warn my leader, it was gone. We both had full throttle but he started to pull away: he had a new plane, I an older re-hab. I hit the mike button and started to speak when a black explosion and a feeling of doom enveloped me.

I was hit and I was blind. No doubt about what I had to do next. To bail out, you had to pull two pins to jettison the canopy. Groping blindly, I could find only one pin, so I reached up to use the regular handle and pulled it back as far as I could. Then, I passed out.

My plane was spinning or spiraling toward the earth, rapidly losing altitude. When I came to, the feeling of doom deepened. I still couldn't see; altitude was running out and so was my life. With deep resignation I said to myself, *"This is it!"*

I felt the weight of my body pressing against shoulder straps and seat belt. *Hey! There's a way out! I must be upside down. All I have to do is pull the release.* I reached for it and lifted. At once, I flipped out of the cockpit; immediately my hand flew to the rip cord handle and pulled hard.

* * * * * * * *

As I sat there in the tranquil setting on the sandy beach by a pool under a waterfall, a feeling of edginess came over me as I looked around. I tried to recall what happened after I pulled the ripcord. It was all a blank. I'll never know exactly what took place; I can only guess.

* * * * * * * *

I found myself blind, hanging in a tree by my parachute harness, feet dangling, the air still, with distant aircraft and gunfire the only sounds. I felt a strong sense of danger, an urgency telling me I was in peril of being strafed from above or sniped at from below. *How close is the enemy?*

In desperation, I grabbed the shroud lines above and yanked them furiously. The thought of being shot seemed far more grim than falling out of a tree. Then I dropped, but no more than five to ten feet. When my feet hit the ground, I reached up at once, grasped the shroud lines, and pulled the chute from the trees. It could be a marker for enemy eyes overhead. Still blind, I unfastened the harness and started running, in fear of strafing from above. Finally, I fell down. My vision began to come back as I lay on my back, propped up on my elbows, the sight in my left eye blurred by an apparent internal hemorrhage. Looking around, I could see that I had fallen in a small rocky stream.

I checked my physical condition. Blood oozed from my broken nose where the left nostril had been split open almost an inch. I had a scalp laceration and painful ribs and wrist on the left side. Standing, I found my legs were okay, thank God.

I thought about my situation. *How would I ever get out of this mess?* It seemed almost hopeless. I was over 200 miles from friendly territory, mostly over water; who knew what lay between the sea and me. *What do I do now?* A feeling of dark despair flooded through me. *I'm in a desperate situation... but there may be a way out. I can hike to the coast, inflate my one-man raft and somehow get to the middle of St. Georges Channel between New Britain and New Ireland. There, I can signal with my mirror, one of the two patrol planes that cover the channel every mission.*

There was a hitch to this plan, however. I didn't have my raft. It was back somewhere in that tangle fastened to my chute along with my jungle kit full of survival gear.

With some vision restored, I quickly hustled back down the slope and soon came across my chute. Hurriedly, I unsnapped the raft pack and the jungle pack, cut off a piece of parachute, and hid the remainder under a bush. A few yards back up the slope I heard, off to the left, a sound like the crackling of a fire. As I moved along peering through the undergrowth, I saw flames. On closer look it dawned on me that it was the remains of my plane sitting there, burning.

Quickly, I turned and hurried off, away from the crushed metal bird that I had escaped at the very last possible second, realizing how lucky I was to be alive.

I worked my way slowly up the slope a few yards, but soon felt drained of energy. Dropping to the ground to rest, I waited for my strength to return and decided to attend to my wounds. I reached for my left shoulder feeling for the first aid kit. It was gone. It probably ripped off during the bail out. I searched through the jungle pack and found one there. It had bandages, sulfa powder, and sulfa pills.

The vision in my left eye was still blurred, indicating possible internal bleeding. I placed a large bandage over the eye to protect it from further damage and to help hold the severed left nostril together, sprinkled sulfa powder on open wounds, and took sulfa pills. There was a minor wound on my left upper arm with a lump in it; I squeezed it and out popped a pea-sized lump of metal. That required only a Band-Aid. Again I started up the slope, away from the distant sounds of battle. The jungle tree canopy was dense overhead; the undergrowth about chest high and not too thick, easy to walk through but the incline drained my energy again. I lay down below the bushes and rested awhile.

Above, there was still the diminishing sounds of aerial combat: bomb bursts, anti-aircraft ack ack, and machine gun fire aloft. Finally, my world was still, except for the exotic sounds of the jungle. Despite the restless urge within me to be on my way, the traumatic experience I just went through had nearly drained my energy.

After a long rest, I tried again. I slowly worked my way up the steep slope and found again that my body was still in a state of utter fatigue. Once more I sank down to the ground below the bushes. It was late afternoon—I hadn't reached the crest and didn't feel like trying. I prepared to spend the night there on the slope.

My injuries, especially the left eye, brought acute pain. Earlier, I had found morphine Syrettes in the first aid kit and decided to use one to get some sleep and be rested for travel next day. I was aware of the pain-killing qualities of morphine, ignorant of its side effects.

I awakened numerous times during the night; once when I thought I felt a very large snake crawling under me, several times by what I took for the shrill whistle of a narrow-gauge locomotive just over the ridge ahead.

In the morning my strength returned, but I was extremely wary of sounds and strange sights all around. I climbed cautiously to the top of the ridge and started down into a gully, but came to a sudden halt when I saw what appeared to be a hut with an old green truck parked next to it. I backed up and proceeded around a knoll to a point where I could look down into the gully from another direction and there I saw it again: the same truck and shed. No people in sight.

Now I was extra cautious. I had no idea where I was in relation to the airfields we had been attacking. With vision in only one eye, I had trouble with depth perception. I turned away from the rim of the gully where the hut lay and headed in a general direction I considered to be away from enemy camps. I had covered only about a fourth of a mile when the first bombing mission of the day began. I peered with hazy vision into the jungle growth a little way off and saw obscure forms of helmeted, dark-skinned soldiers. I dropped to the ground and lay still for a long time, waiting, listening, hearing only fading sounds of planes in formation. Every day our planes came over in three waves: at eleven, twelve, and thirteen hundred. Just before the second wave of our bombers started, I ventured another look. I raised my head slowly, opened my eyes, and peered through the leaves in the direction I had seen the soldiers. To my horror I saw one of the soldiers leaning on his rifle but the one next to him was pointing his directly at me! I dropped my head back to the ground and waited breathlessly, expecting to hear footsteps coming toward me. I waited and listened.

My God, did I really see a gun pointing at me? I don't want to look again. He might shoot. I'll just lie here and wait with my eyes closed. I wonder why he doesn't fire. I could wave a white cloth. I have the piece from my parachute. It's in the jungle pack on my back and I'm lying on it, but I don't want to move. I'll just lie here and keep still and hope they didn't really see me. Maybe they'll go away after the bombing stops. The bomb bursts sound muffled; must be five, seven miles away, interspersed with aerial gunfire and ack ack.

These thoughts passed through my mind as I lay there, frozen in position, waiting for the last wave of our attack planes to pass over. I felt as if my body were sinking lower into the ground. *Was this wishful thinking?*

With the waiting, my mind began to wander. I thought about my family: *are my cousins in the service in dangerous places like this? Or my uncles? Have any been killed? In earlier wars? Oh yeah, Mom's Uncle Guy, my middle namesake, lost his life in Cuba during the Spanish-American War. He was quite young in 1898.*

Dad's Uncle Will was shot by a man in uniform, not a soldier, not in a war, but in a forest, by a constable. Will served on a posse looking for two bank robbers. He and another posse member were resting on a log when the Constable came upon them in the woods; surprised them. Uncle Will, hard of hearing, jumped up, took a gunshot in the leg, and fell. His partner whipped out a white kerchief and waved it, furiously, averting further gunfire.

One posse member had shot another. The gunshot wound had a difficult healing, with bone loss that resulted in a shortened leg. For the rest of his life Uncle Will was hampered by two handicaps, both closely related by cause and effect. I can still see him limping across the suspension footbridge that he erected across Lewis River to the cabin he built in his retirement years, in spite of his handicap. The bridge would sway in a hippity-hop way when he used it.

I lay there, frozen in the same position until all the bombing stopped for the day. At one point I thought I felt a grass stem tickling my ankle inside my pant leg. I kept my eyes shut; *they're toying with me, why don't they just take me prisoner and get it over with?*

Slowly, I opened my eye a slit. There was no one there! But I was still wary. Then I heard what sounded like a small pebble skipping through the brush above and landing on the ground nearby. Another missile came in from a different direction, and then another. Now I was really getting annoyed. I got up on my knees and pulled out my forty-five. I stood up, looking around in all directions. I was totally alone. But the little pebbles were still coming in and I started pointing my handgun in the direction they seemed to come from. Now weirdness had really set in. *I'd rather be taken prisoner than declared mad.* I suddenly realized I had been pointing my gun, like a demented fool, at some imaginary beings. I would have seen the humor in the situation if I hadn't wasted most of the day.

Looking around, I could understand that with one eye, broad leaves on trees would look like helmets. There were no soldiers hiding in the jungle aiming at me; the hut and green truck I had seen back there in the gully in the morning were merely constructs of a drug-driven imagination.

Inklings of pain from my wounds were returning. I was downright disgusted and hungry! I realized I hadn t eaten in almost thirty-six hours and that creature needs must supplant paralyzing fear. And I needed to get moving even though the day had faded to late afternoon.

I found three cans of pemmican and a chocolate bar in the survival pack. I ate some of the chocolate and half a can of pemmican, took some sulfa tabs with water from my canteen, slung the pack to my back, picked up the raft pack like a briefcase, and went crashing off through the jungle toward the distant sea.

At first I moved along quite well but as darkness came, one-eyed vision became a definite handicap. Shining down through the trees, the moonlight gave the illusion of a smooth blanket of snow when it was really a tangle of logs, rocks, and low-lying plants. I had to pick my way very carefully through this obstacle course while disregarding what I saw.

I came across a meandering brook that seemed to be going my way. Instinctively, I started to follow it. For about a half-hour, I made some progress when the creek started getting noisier, making a sudden right turn and plunging down the hillside. I almost went with it, but I grabbed a branch and pulled myself back up, retraced my steps, and gave up on the faithless creek.

As I laboriously made my way in what I guessed to be the southeast, I realized I was getting tired and decide to just flop on the ground and rest. With the piece of parachute to wrap around me for warmth, I actually slept for about an hour until a drenching rain awakened me. It soon had the parachute sopping wet, as well as my clothing. I tossed the dripping cloth aside and resumed sloshing through the jungle tangle.

At dawn, sunlight filtering down through the trees warmed me and made me feel sleepy. I lay down again on the jungle floor, my new bed, and dozed off for a while.

When the morning sun streamed down through the jungle canopy, I opened my eyes slowly. I lay there awhile, thinking about direction. The sun coming up in the east showed the way I wanted to go. Then, I remembered some things my dad had taught me about finding your way in the forest and the family story about Dad and Uncle Will getting lost in a cloudburst, then finding their way out.

Uncle Will limped along the forest floor, sloshing through what seemed to be a pond, trees all around, small jack pine, in a heavy downpour with water two or three inches deep. Dad sloshed along behind, glancing around. They slowed, and then stopped, peered around through the deluge. Almost to their ankles in the shallow water, they looked at each other.

"I think we've been going in a big circle," Dad said, "We were here quite awhile ago. I recognize this fallen log here."

"Yeah," replied Uncle Will, "All this water has wiped out the trail. No blaze marks anywhere... can't see more than fifteen feet ahead in this rain."

Both of these men knew how to travel in the wilderness under ordinary conditions without getting lost. They looked for landmarks to get their bearings and if the forest was too thick to see distant peaks, ridges, and gaps, they used the sun to keep a straight course.

Those were no ordinary conditions that day. They were in a cloudburst, which means actually in the clouds when a tremendous volume of rain is being released. A layer of water obliterated the trail. The terrain was flat lava bed covered with small pine trees that all looked alike; visibility was almost zero at times. Uncle Will had been leading and with his one short leg had inadvertently taken them on a wide circle back to a place they'd already been. They couldn't get out of there until they found a way to hike in a straight line.

They were on their way home from a successful fishing trip to a remote lake on the slopes of Mount St. Helens in the Pacific Northwest. If they didn't get home that night, two families would worry about them. They had to reach Lake Merrill first, which was only about five miles away. But *which* way? And, how would they stay a straight course?

Dad looked down. He's a shy person, gentle, but self-effacing. This day it stood him in good stead. Focusing on the water swirling around his feet, he noticed something. *The water is flowing. It's flowing downhill. This means if we follow the flow it could take us to the river that runs past Lake Merrill. There, we could find the trail to the lake or smoke from campfires, familiar landmarks.* He pointed to the water as he explained to Uncle Will what he discovered and they agreed that it would be better to follow the flow than to risk continuing going in circles.

With renewed energy and elevated spirits they struck out in the direction the flowing water led them. They were elated. It was still quite a hike but they felt they were making progress. They were still going to be overdue and might not be able to drive on home that night because of the storm. After two or three miles across the lava bed, the terrain started to slope sharply and they found themselves at the bank of a roaring stream.

The downpour continued, but they found a trail along the bank, which they followed, and soon it crossed another trail, which had a sign pointing to the lake. Soon they were at the campground. What a relief. The plan worked. It became a family maxim: "Follow the flow."

"*Follow the flow*," I was sleepily reciting the lesson. Slowly it came to me. *Hey it applies. I can follow the flow! If I travel in a straight line southeast until I find a stream that is flowing in the right direction, I can follow it to the sea...* Then, remembering that little roaring brook I had followed in the dark, I thought, *but not at night.*

I finished the can of pemmican, ate more of the chocolate, and spent a few minutes getting better acquainted with my jungle pack. I found a compass, matches, signal mirror, fishing gear, flare gun, and a dye marker. Most of this was going to be useful once I'd launched into St. Georges Channel.

I took more sulfa pills with plenty of water. My water supply was getting low and I looked around for tree vines. We learned in survival school that if we cut the vines trailing down from the tree canopy, clean filtered water would drip from them. I tried it and it worked surprisingly well. I filled one of my canteens from just one of the larger vines. I also remembered from survival instruction that you can get heart-of-palm, a gourmet delicacy, from baby palm trees. Later that morning I found one in a sunny glade and rejoiced at the good fortune of not only finding food but finding a delicacy. As I cut into the base of the small trunk, I salivated in anticipation of a little feast. But alas, I found no tender heart in that palm; just hard, dry wood all the way through. Later I learned that only coco palms have that tender heart.

I started off again in the direction of the morning sun, thinking it to be about southeast (I later learned that below the equator at that time of year the sun rises due east). Being at a fairly high altitude and with the undergrowth fairly sparse, I made good time.

At a point where the terrain started to slope down in a good direction and the overhead canopy especially dense and dark, I came upon a wide, well beaten path, which I crossed hastily. Assuming it was a native path, possibly between villages, I put it behind me, quickly.

Late in the afternoon, I came across another bouncing, gurgling creek heading my way and I decided to follow it for direction and comfort. The terrain sloped gently ahead and the walking was easy. Soon the banks on both sides steepened, so I started wading in the shallow water since the going seemed more even and with a more solid footing. As it became darker and some of the little waterfalls got higher, the going was treacherous and I decided to stop for the night. I'd already had a close call with the wild meanderings of a stream in the dark.

Resting on a large flat boulder, I ate a half-can of pemmican, feeling the need to ration my small store of food.

I stretched out on the sloping boulder, feet downhill toward the stream, with the flat jungle pack on my back and the raft pack at my seat. Almost immediately I heard and felt mosquitoes buzzing around my head and recalled the threat of malaria in those jungles. I remembered seeing a mosquito net in the survival kit, so I got that out and put it on and felt immediate relief from the little suckers. But soon they buzzed around my hands, and found tender places. I did my best to tuck my hands under my shirt. Finally, I fell into a fitful sleep.

It was a hard night on the rock but I slept enough to be energized when dawn came and I started off down the bubbling creek. The banks were getting steeper. Very soon I could again see why it was a bad idea to navigate unknown waters after dark. My little friendly brook was about to become a full-blown cataract as it flew off into space before me.

I turned around, retraced my steps, then climbed an embankment, walking along the crest of a ridge until I came to an incline covered with loose shale. There were no trees growing on this slope and I was high enough to see over the jungle ahead. I still could not see the coastline, just treetops stretching out to the horizon. I felt disappointed by this, but heartened by the sight of a small river in the valley below. Following the flow seemed to be working: the stream was flowing in a good general direction.

Cautiously, I slid down the shale-covered slope, wary that a stream below could bring humans. I found a most delightful, clear, sparkling little river at the bottom. Off to the right was the waterfall made by the creek that I had been following up above. The river was shallow, with a bed of rocks, boulders, sandy stretches, shallow rapids and riffles. I soon found that it also meandered through the jungle like a snake. At the first bend I was surprised and shaken at coming across a small native shelter made of poles and a grass roof. Dried bones hung from the roof. No one appeared. *Is this a sign that there are going to be a lot of natives along the river?* That was my fear. I moved along cautiously, trying to see ahead while still keeping an eye on the large rocks I was jumping on. As I hopped boulders, waded through shallow riffles, and walked across gravel benches, I had a strange feeling of comfort and security in spite of the obvious possible perils. It reminded me of home. The Pacific Northwest, the rivers and hills and forests where I grew up, were almost like these. Only the trees were different.

When I stopped to rest and have some rations, pills, and water, I thought about fishing. On fishing trips with my Dad, he would be in his element. With a rod in his hand, he could fish for hours without tiring. I had no patience for fishing but I loved boulder hopping, dashing up and down banks and splashing in the water. The environment was so beautiful, I loved just being there and being with him.

I liked to think Dad's profession was "fly fisherman." When he fished in Pacific Northwest forest streams, he was in heaven. A quiet, shy man, he had a special look on his face, a perpetual half smile when casting the fly. Fishing made it all worthwhile, the hard work during hard times, the beginning of the "Great Depression." He made a living as a foundryman. It was hard bending-over-work, making sand molds for cast iron products. Starting in a cold foundry in the morning, working with cold black sand that permeated his clothes, smudged his face, and left dark creases in his hands. As an artisan he used his bare hands and his great knowledge of the craft. He was a skillful craftsman.

By noon the furnaces revved up and, summer or winter, it got hotter by the minute. When the time came to pour the molten metal, he stood by his molds with a rod in his hand to skim off the slag, as the big ladles of glowing molten iron were hand carried by two men to his molds. He stood there skimming until it was done and the whole foundry steamed from the hot, damp sand. He used to joke about the ladles brimming with Satan's ice cream. When the iron had solidified, the men performed what they called the "shake out." That's when it got really steamy and smelly. The sand was shaken from the flasks along with the hot solidified iron castings. I can still smell the vapors and see them rising from this apparent chaos of overturned flasks, sand, and weird metal forms.

That was Dad's craft, but not his profession. His profession was fly-fishing and he was one of the best. His office was on the beautiful streams of the Pacific Northwest. Here, this quiet, dignified, handsome man waded the riffles, skirted deep pools, boulder hopping, casting the fly in just the right places and catching nice rainbow trout. Not a bad profession: catching rainbows!

While he engaged in his skillful activities on the river, I practiced my trade as a boy on the river. I was a boulder hopper too, not to find good fishing spots, but for the sheer joy of it. Clear, rippling water swirled around and over rocks and boulders, sunlight sparkled off the cool liquid, under the shade of overhanging branches in our little corner of Eden.

All morning as Dad moved along the stream, I was close by, seldom out of sight, dashing up and down banks, jumping over logs, sliding down huge boulders, leaping small chasms, swinging on limbs, until one day, I grasped a too small, too dead branch. It snapped and I found myself suddenly down across large rocks, lying in water with scraped knees. Stunned, I looked up and said, almost aloud, *What am I doing down here?*

I sat there by the river, looking up and asking myself, *what am I doing down here? I was up there not long ago, now I'm down here just like long ago in my boyhood, on the rocks.*

    Right down there close to the water I could see fish. They looked like those trout we caught on the Lewis River in Washington. Searching through my pack for fishing gear I found a line and hook with pieces of dried bacon for bait. In the water, there were more fish and crayfish swimming around the rocks. Remembering using crayfish for bait, I also recalled how much time and patience it takes to do all that and how little I had of either. I didn't want to while away valuable time on a doubtful outcome. So, I opted for another half can of pemmican, then picked up my gear and moved off again around never ending bends and stretches of that beautiful stream.

    In its twisting and turning it seemed to be trending northeastward, closer to the coast but also closer to enemy airfields. In the middle of the day when the bombing started, the sounds of war had gotten louder, but no planes had come into view over the river. Constantly alert to my surroundings, I scanned the riverbanks, watching for movement, listening. The afternoon was uneventful except for the beautiful vistas that opened up as I cautiously rounded each of the river's many turns. As the sky grew dark I started looking for a likely place to spend the night. At last I found a soft sandy spot behind some low bushes and settled down for a more comfortable night. As I gazed up into the deep night sky, I thought about that river that reminded me of home. There was something familiar about the sounds and the smells. Something was coming back. It was the smell. It reminded me of the river willows that grew along the Columbia River when I was a boy growing up there.

    When we thought of a place to cool off on a hot summer day in Vancouver Washington, it was usually the river, the closest river, the mighty Columbia right there at the foot of Main street. My friends and I would just start walking, shirtless, in our shorts. We were completely unencumbered, without even towels, free as the birds. At the foot of Main Street stood the Interstate Bridge. We took that mile-long span to our destination on Hayden Island. Jantzen Beach Amusement Park lay on the right, our destination on the left. Through a grove of cottonwoods and river willows, we followed a network of trails to the edge of the slough on the far side of the island. There we found a little cove where the water circulated so slowly that it had time to heat up in the summer sun. It was our private, solar heated swimming hole where we could go skinny dipping in total seclusion. We would strip down to nakedness and run shrieking into the warm water. It was such a feeling of joy and freedom. The main attraction on that island was not at Jantzen Beach. Not too many people knew that. Thank God.

I slipped off to sleep with a smile on my face. Some of the stress of my ordeal had been relieved. I slept on through the night to the soothing tones of the flowing river alongside my sandy bed.

CHAPTER 2

WORKING MY WAY OUT

Then along came February 14, which started out so badly, but turned out to be St. Valentine's Day, thanks to my loved one, who showed up in spirit to hearten me. My clothes, lying on the beach, were still damp but with the sun out they would dry on me. I put them on and away I went with renewed energy.

It was another pleasant, uneventful afternoon on the same old winding snake-of-a-river. It seemed to be getting a little deeper in places. Sounds of an airplane were nearby but it didn't fly over. Late afternoon I looked for a good place to sleep on my fourth night in the wild. I found some taro root back a little way and I was eager to have a hot meal. I looked for another sandy beach but I found none and it was getting late, so I settled for a gravelly bed amongst rocks and logs.

I gathered some pieces of wood, made kindling with my knife, and started a fire. In survival school we learned that taro root must be cooked or severely beaten. I opted for cooking because I wanted a hot meal. I wrapped the piece of root, about the size of a half of a potato, in a broad leaf, built the fire up around it, and let it steam for about fifteen minutes. My mouth was watering, readying itself for a small tasty meal, anticipating a warm succulent morsel. But, ugh, it was like a mouthful of pins! I spit it out and discarded the remainder. I just didn't give it enough time in the fire to break down the crystals.

I finished off the can of pemmican I started yesterday, then got ready to settle down on my rocky bed. My clothes were still a little damp and I felt cold, glad I had the fire. Some of the rocks near the fire, about the shape and size of potatoes, were hot. I tucked them in around my shirt and increased the comfort level, somewhat.

Soon I was in dreamland, thought I was back with my squadron; Doc Donnely, our flight surgeon, was joshing me about the rocks in my shirt. He said, "Now you are the original Hot Rock Kid."

It was a fitful night's sleep but I rested enough to recharge my "batteries" and at dawn's first light, I was eager to be on my way. The river had flattened and broadened, running quite still. As midday approached, I found myself wading waist-deep in places and it was getting uncomfortable. As I rounded a bend I saw a sandy beach with a small banana tree sporting a bunch of its bright yellow fruit. I hastened over to the little tree, lopped off the bunch with my knife and prepared for a feast. Grasping one of the bananas in my hand, I started to peel it in the usual manner but found it unpeelable. I took out my knife to slice it and found it hard all the way through, like a cucumber. I learned later that these were not bananas as we know them, but plantains, which require cooking to make them edible. *Fine thing. I thought the jungle would be full of ready-to-eat delectables.*

I looked down the river beyond the sandy beach and noticed that the edges had disappeared. *The stream goes all the way to the bushes.* There were no banks in sight and the water looked deep. I decided to inflate the rubber raft I had been carrying all the time. It was just getting too difficult to wade anymore.

The pack holding the raft also contained an attached CO_2 cylinder that could be used to quickly inflate the small craft, but I saved that handy aid in case I needed it in a hurry later. I would blow it up by mouth.

Lying down next to the spread-out raft pack, I looked around. From this vantage point I could see, over low-lying bushes, in all directions without being seen. I could relax as I blew into the long stem. With slow, deliberate breaths I slowly inflated it. It took quite awhile, but going slow conserved energy and I needed that. Lying there, kind of relaxed, blowing into this tube reminded me of something I had done before. We used to blow up our footballs this way. I remembered the first football I owned. Uncle Kelly gave it to me for Christmas in 1925.

Uncle Kelly came to visit over the holidays. I felt the excitement of having him there. He was a college student at Washington State at Pullman and he brought me presents from there. One was a rooter's cap, the kind that is shaped like a sailor hat, except this one was red and gray with a "W" on the front. The other gift was a treasure, a real full-size college football straight from the playing fields on campus.

He told us the story of how he happened to come by that ball: his course in forestry took him on many field trips and one particular day found him in the boondocks with surveying equipment next to the football practice field. He had his head down over the eyepiece when he heard a plop. He looked around for what he suspected would be a football somewhere in the bushes. Sure enough, he spotted it hung up in a low shrub. It had flown over a high hedge from the direction of the sounds of football. Kelly thought, *if they come looking for it I'll tell 'em where it is.* He was that kind of guy. But he knew that if they forgot about it, he would help himself to it by smuggling it out under his jacket. That was his prankster side.

Back then, footballs were fatter and rounder than they are now and it would be an obvious bulk under his jacket. But, in those days footballs were more quickly and easily deflated. When his class was over, the ball was still there. So he simply unlaced the strings, pulled out the stem, let the air out and tucked it away out of sight. He walked away, unseen, with my ball. He brought it along with the rooter's cap all the way across the state of Washington just for little old me. I felt real proud and special.

Santa brought me a shiny red wagon that same Christmas and it was a thrill to me when Uncle Kelly took me for long rides all over town. We went to different places every day. I could hardly wait to hit the streets and byways with my uncle. He was special.

Finally, I came out of my reverie, realized the raft was fully inflated and that it was time to get back on the river in a new mode of travel. Taking a quick look up and downstream, I loaded my gear and launched the raft, on my way again down the widening stream, floating gently through the deep places; carrying the raft over shallows. Delighted by the stretches of deeper water, I felt that it meant I was getting closer to the coast. However, it was a little disconcerting to be floating in broad daylight in enemy territory, in a bright yellow one-man raft.

I floated and portaged for the remaining daylight and on into dusk. Darkness fell as I floated quietly through a tranquil stretch of river, with small fish jumping and smacking the surface. I heard one plop into the inside of the boat and I could feel it wriggling around in the puddle of water there.

I found another sandy beach, still warm from the afternoon's sunny rays. Entranced by the ambience of this little beach, I decided to make it my bed for the night. I pulled the boat out of the water and emptied it, forgetting the fish. Another creature had been spared by my ineptness as hunter and fisherman. So I settled for another meager ration of pemmican and bedded down for my sixth night in the jungles of New Britain, hoping to find myself at its shore tomorrow.

CHAPTER 3

THE GETAWAY

The next morning I was up and moving before sunrise, with high hopes of an early arrival at the coast. I still encountered stretches of shallow, fast moving water; the pools were widening and deepening and there were fewer times now that I had to get out and carry my raft over the shallows. In the middle of the day I neared a bend in the river where the water ran fast and a bit deeper. I decided to stay aboard and run the rapids, focusing on the water for sharp snags, below or near the surface, that could puncture my craft.

It was not until I had gotten through the roughest of the rapids that, lifting my gaze to the banks, I was horrified at what I saw. Dead ahead above the shoreline was a coconut grove. I had reached a place of danger. *There must be people. I gotta get outta here!* I pulled over to the shore immediately and dragged the boat across a wide, well-beaten path and into some high grass where I laid low and listened.

It was almost mid-day, maybe close to siesta time for the locals. I rose up on my knees and looked around. That pathway, so close, made me nervous. I needed to find a better hiding place. Flies, big ones, were buzzing around my head, even nipping at my hands and face. *I can't stay here.* I stood slowly, gazing warily around. No one in sight. I could see that the river made a turn, downstream, to the left.

I left the raft hidden in the grass, waded across to the other side, then reconnoitered through the bushes to the place where the river went, to see if it was clear on downstream. There didn't seem to be any danger down there so I returned to my raft, took another look around, then picked up the boat and launched myself quickly into the stream. I let it take me to the sharp bend amid tall reeds at the foot of a steep bank just below the plantation. Now, with the bright yellow raft well hidden, I climbed up into the grove. There was a rise off to the east and I walked that way in hopes of seeing the waters of St. Georges Channel. There was too much vegetation in the way, but I knew the ocean was there. I could smell it.

It looked as if I'd be taking to the sea very soon and, fresh out of rations, I looked around on the ground. Intent on finding coconuts, I found three and took them down to the raft. Looking around cautiously, I launched again into the current.

All went well at first, until I completed the turn and to my alarm, found myself drifting slowly past a pleasant little beach with bare footprints all over it. I was probably gliding through the bathtub of the enemy, if not their rec center. I paddled furiously with my hands to the opposite shore where I spied a dense bamboo thicket. I clambered up the bank, pulled my raft quickly up with me to sit and rest and wonder at my good fortune in avoiding danger.

A little later I tried to open a coconut. I drilled holes in the ends with the point of my knife to get to the milk, and then finally pried it open. I eagerly chomped on the white meat for awhile, thinking I would eat the whole thing, but soon found I had a limited capacity and gave it up halfway through.

Still later, I heard pounding off in the distance like the sounds of someone nailing and building something. When it stopped I ventured out of the thicket to investigate further downstream. Only fifty to seventy yards I found that my little river entered into a larger river running parallel with the coastline. A large tree spread long branches low over the water at the corner. This would be a good place to hide at the beginning of darkness.

I waited until dusk then pushed away from shore again, and gliding silently to the big tree, reached up and grasped a branch to hold me while I awaited full darkness. When it came, I let go and drifted slowly into the current of the larger river. The dark shoreline showed on either side. I saw a darker form on the nearer starboard side. There were vertical columns topped with a platform. Looked like a wharf, probably where they loaded coconuts on barges.

Drifting on past the wharf and around a wide bend, I felt the slightest rise and fall of the first waves from the sea as the shore receded slowly on both sides. More relief came with the passage of each minute down the widening stream, until something appeared on the obscure horizon. A dark form spanned the river mouth. Coming closer, I could see that it was a bridge made of a long string of pilings with planks on top. I worried that this might be an impassable barrier. I saw that the space between pilings was wide enough... *But what about barbed wire in that space? I just have to trust to luck it's clear.*

I closed my eyes as my raft and I passed on through the space under the bridge without hindrance of any kind, but just as we cleared the structure I heard a shout through the rising wind off in the distance. It was in a foreign tongue, decidedly Japanese. *Did he see me? I just have to trust to luck again. I've been doing a lot of that lately. I also need to pray I'll reach a safe distance from shore by morning.*

One shout was all I heard. No searchlights. The bridge faded out of sight behind. The waves got higher as I reached farther into the bay. A nice tailwind joined the river current to hasten my escape. The dark unknown ahead felt better than what I knew lurked behind. Morning would tell how well my escape had succeeded, how far from shore I had drifted. I was pretty much at the mercy of nature's forces here: the river current, the movement of the sea, but most of all, the wind.

I had no oars, but even those would have little effect in face of a strong breeze. The inflatable raft reacted to the wind much like a sailboat. A large portion of the air filled ring lay above the surface. I did have a small paddle that strapped to the hand, small sailcloth and a sea anchor, but I wanted to go with the wind and current as long as it continued away from land. The sea anchor was needed only to keep moving bow first. I just drifted away from land all night, sideways, rocking gently, dozing.

As dawn showed its faintest glow in the east it revealed the obscure profile of New Ireland, the other island that formed the dark channel I rode on that seventh day of my ordeal. The dark form of New Britain rose behind me. I expected that, but how far from shore I had drifted, I would soon know as the sky slowly brightened.

In the water around me several objects floated. It looked like garbage: an old crate, pieces of fruit, what an old sailor might call flotsam and jetsam. I could see, as I floated on the fringe of the debris, a sudden change in the color of the water: on the shore side, a lighter hue; but to the seaward, a deep, dark value. Searching, shoreward with my one eyed gaze, I looked for signs of possible danger. It was still too dark to tell how far I had come; only the dark forms of distant mountains silhouetted against the lighter sky to the west were in view.

I turned again to the east. The sky had lightened a little and on the surface something I saw gave me a twinge of fright: a large dark form much like a dugout canoe with a slim figure standing motionless at one end. I thought it must be a native. I watched for any kind of movement, but the figure stood still. *Is he watching me? Where did he come from?* The only motion came from the rise and fall of the waves. Straining my one good eye in the semi-darkness, I tried to see if it was what I thought: a native in a dugout?

As I strived to focus on this apparition in the growing dawn, something caught my eye on the brightening horizon: just off to the right: a small stream of dark smoke trailed off behind some unknown craft just coming up over the horizon.

It engaged my whole attention as I watched it approach. Its course was not right at me but off to my right and headed toward the bay I just departed. Its profile was short and blocky. I thought, tug boat at first. Black smoke continued to pour from its stack as it hastened toward shore. Then it changed course to the northward, parallel with the coast, now clearly in view. It had guns fore and aft on a very low deck with a square blocky superstructure in the center of which was, clearly visible, a white square with a red dot.

An alarm went off in my head, *Hey, that's an enemy submarine! Surfaced!*

I immediately reached for the sailcloth, turned the blue side up, and pulled it over the raft. At this vantage point I really saw the nature of my surroundings in the light of dawn. This fringe between two different kinds of water was a resting place for flotsam and jetsam, the place where muddy green river water met the deep blue of the sea. A floating tree trunk; a tree with one root sticking up, looked a hell of a lot like a native standing in his dugout.

I peeked out from the edge of my sailcloth toward shore. The sub had dropped over the horizon ahead of its smoky tail. The shoreline was out of sight over the surface but the tops of the coco palms were still in view. I may have been saved from detection by all that stuff floating around: old pieces of fruit, a crate, flotsam, jetsam and a tree—a bit of personal disguise. *Is this my good luck at work again? With wind and current and camouflage?*

I drifted to the southeast and as the plantation faded away and the middle of the day approached, I reckoned I had reached a position between five and seven miles from the tip of the Gazelle Peninsula. I searched the eastern skies for the first wave of Allied bombers of the day, for I knew the two search planes would be close behind, flying low over the channel.

I must bring out the signal mirror from my survival kit and start practicing again. I learned the technique in the survival class back at Santo but it would be much more difficult in a rocking boat. There had to be a way to hold the mirror steady. Finally, I found it by bracing my elbow on my knee. I was sure it would work.

Then it came! I heard the distant drone of many aircraft in formation. *I must be alert and continue to search the skies, look for that twin engine, twin finned Vega Ventura, the PV-1 that skims the channel surface in search of downed flyers and enemy submarines.*

With hope rising, I had faith that my plan, made seven days before, might be nearing fruition, that the mirror would work, I would be seen, and my rescue was imminent. The stress would soon be over.

Shortly after the second wave of our planes had passed over, I saw it: the dark, side profile of the Ventura I'd been waiting for, way off in the distance, silhouetted against the light blue hills of New Ireland. I pulled out the signal device. Through the hole in the middle of the round mirror I could see the side of my target. I struck it with slashes of sunbeams. Almost at once it changed direction and headed directly toward me; then the small speck of a target started to grow and grow, as if fed by the sunbeams streaming from my hand.

Quickly it was upon me, off to my left as it flew over. It had New Zealand markings on wing and fuselage. I got a glimpse of an airman in the nose bubble. I saw a waving hand. I felt joyful, grateful, and confident that I would soon be in safe hands. They proved it at once as they made one counter clockwise circle and dropped a canteen a few feet down wind from my raft. I floated directly to it, within easy reach and picked it up. *Boy! These guys are really good.*

Another circle or two and they dropped a bundle in the same relative position. The wind blew me directly to it. I simply reached down and hoisted it in, thinking: *This is probably rations. Does this mean I may have to wait a few more hours before they send the rescue plane?*

That question was answered immediately when my benefactors came out of their circling mode and took off in a southeasterly direction, leaving me to pursue my own passive course in the same direction at the whim of a helpful wind blowing my way.
Why, why didn't they send for "Dumbo"? Am I still too close to shore? Guess what I need now is patience and a favorable breeze. And... lunch!

There was that big bundle 'specially delivered by air and I hadn't even opened it. Turning it around and around looking for the zipper, I finally opened it and found a big olive drab, wax-covered box. A large, simple label in black letters read: PACKED BY THE CRACKER JACK CORPORATION.

My God! It even has my name on it and one of my favorite childhood goodies. I wonder if there are any prizes?

Slipping the knife my Dad made for me out of its sheath, I slit the top of the box open and could see at once it was not Cracker Jacks, just cans. I pulled one out in anticipation of my first real meal in a week. The can looked rather familiar. I turned it around to read the label. It said, PEMMICAN, the same stuff I had in my jungle kit.

Oh, well. At least it's not going to be too much of a shock to my digestive system.

After consuming one can (I could only handle a single can), I looked around at the empty sea, the two islands that formed this channel, and estimated that I was still closer to New Britain, but given the prevailing wind and current I could reach mid-channel some time that day. It would feel safer then.

The wind picked up a little, and so did the whitecaps and the swaying of my craft. I must have been dashing along at a two to five knot clip and could be well out of danger sometime in the afternoon. I had my fingers crossed. *They could still get me out of here today*, I thought.

Warm rays of the mid-afternoon sun along with the rhythmic motion of my boat put me in a sleepy stupor, but my drooping eyelids snapped open and my head came up when the distant hum of a single engine aircraft reached my ears.

Aha, they have returned. I searched the skies off toward New Ireland.

Then I saw it. *But, it's a single engine low-wing monoplane. This is unusual. I expected the twin engine Ventura or Dumbo.* It was flying low, almost directly overhead at about five hundred feet. I focused on its wings. *Oh, oh. Red meatballs!* It passed on over. I looked for my sailcloth, but by the time I got it out, danger had passed. He may have been looking for me but the afternoon sun must have created blinding sparkles off the surface of a sea already covered with whitecaps. The plane flew on, disappearing in the blinding rays of the afternoon sun. Again, I gave thanks to my fantastic good luck. Or was it?

Just before sundown, I spotted another plane coming toward me from the southeast. I recognized the head-on profile of the friendly Ventura, like the one that circled me earlier. I whipped out my mirror and got his attention at once. He came roaring in at top speed, went right over the top of me and kept on going toward Cape Gazzelle, at the northeast tip of New Britain. Then made a broad, sweeping right turn that took him south of me. He straightened up and headed toward the bay I had left the night before. I saw geysers of seawater rise up in his wake, followed a few seconds later by a rumble in the water below. The plane must have dropped a depth charge on a submarine that started early on its nightly mission, or the pilot may have feinted an attack to draw attention away from me. They did not want the enemy to know their real reason for coming out here: to plot my new position.

The PV-1 flew off just as the sun set, and I prepared for another night on the water. The wind had picked up, making occasional white caps that broke into the boat, but leaving me sitting in a perpetual pool of seawater. I tried bailing with the canvas bucket that doubled as a sea anchor, but the effort drained my energy and seemed almost futile. I had no fear of the boat sinking because of the inherent buoyancy of an inflatable raft, but sitting in the water all the time, I found quite uncomfortable. I tucked the sailcloth along the windward side to help ward off some of the splashing that tumbled in.

Sometime after dusk I heard planes flying overhead. I couldn't see a thing up there on that starless, moonless night, probably overcast. I learned later that was the night the enemy pulled many of its planes out of the area to help defend their base at Truk in the Central Pacific, which had been attacked by our Navy that morning.

When the last of the sounds of flying aircraft died away, I heard only the whistle of the wind and the splashing of waves around the boat. I dozed intermittently. The water inside the tiny craft had gotten deeper and I tried bailing with my canteen. I lay on my side with one hand outstretched, holding the canteen under water, listening until the "in-glugging" stopped, then effortlessly turned it over, waiting until the "out-glugging" ceased. This seemed even more futile than the bucket I used earlier, but I lost less energy and it worked—for a while—until another wave hit. I simply went on doing that mindless thing to occupy time, to maintain my morale.

At a point while going through those repetitive motions, I heard a roar just outside the sailcloth. Raising its lower edge, I peered out as a dark form slipped past. I smelled diesel smoke, and I knew at once that a surfaced submarine ploughed past only inches away—presumably the same one I saw the previous morning. In seconds it had passed and my boat bobbed and swayed as I rode the surging wake. It happened so fast I didn't have time to be frightened. That was about as close as you get without being seen or even rammed. *What luck! Or was it?*

I resumed the bailing for awhile then succumbed to dream time. The waves seemed to be lessening and the ride more conducive to sleep. The dream was strange. My whole squadron was there, each guy sitting in a one-man raft. I was completely surrounded by my comrades. *Is this wishful dreaming?*

In the light of dawn, the dark profile of New Ireland appeared.

In fact, I was approaching the southern tip of that island, Cape St. George. I had covered a remarkable amount of water in the past thirty-six hours.

The sun came up and warmed me after a little slow bailing followed by some dozing in the sun, waiting for the first planes to fly over. I felt a sense of relief; it wasn't over yet, but the third part of my escape plan, the actual pickup, could be only about three hours away. The nightmare, the apprehension, the stress started to fade.

And then they came. First I heard the familiar drone of planes in formation, and then I saw the specks that were our fighters and bombers high overhead. Probably I was being premature and overeager, but I just knew I would be out of there that day. A lighthearted feeling cheered me. I already had my signal mirror in hand an hour before the search plane's arrival on the scene. Then I found myself practicing on the overhead formations; first on a flight of Dauntless dive bombers, then on some Tomahawk fighters, Corsairs, Hellcats, Mitchell bombers. It felt like a celebration. I didn't expect them to break formation for my sake. I was just acknowledging their presence, with the flash of a mirror telling them my happiness.

A half-hour later came the happiest moment yet. With the next wave of bombers there appeared the two low-flying patrol planes over the horizon. *There they are! My mirror is at the ready, the flare gun and the dye marker are close by, this has to be my day.* I flashed a sunbeam at the nearest plane and it immediately vectored toward me. I dumped dye marker liquid overboard just to make a bigger target. The PV-1 circled around my raft at once and I waved furiously to him in gratitude and joy.

Then through the sounds of the nearby Ventura and the distant formations came the familiar, shrill whistle of a Corsair. I searched the sky and then I saw it: the well known head-on profile of the gull-winged, bird-like F4U-2, as it dove straight toward me. I thrilled at the sight and sound of this, my favorite war bird, the one I loved to fly. The pilot came roaring in, throttled back, joined the traffic circle, and pulled back his canopy. I could see him clearly as he waved to me. I had a lot of company now. Rescue was imminent; I quivered with anticipation.

It took about an hour for the PBY flying boat to come up from the Treasuries. Now I patiently awaited the really big moment. Sitting in the late morning sun in a small rocking boat would normally put me to sleep, even in enemy territory, but not that morning. Adrenaline was flowing. I wanted to remember all the details to tell my children and my grandchildren.

Then it came! The familiar profile of Dumbo (the name pilots gave to the Navy's flying boat with the oversized wings) slid up from the horizon, the happiest sight of all, and as it drew nearer I could see it had a fighter escort of nine Corsairs flying alongside. What a picture that was.

As Dumbo glided in, my two protectors pulled away to make room for the rescue. The seaplane flew over low on the first pass, dropped a smoke bomb for wind direction then circled to the left, came around and let down to make the most breath grabbing landing I had ever seen. As the water sprayed up around the plane, I rejoiced. It was like a celebration. It had the festive air of rituals in New York harbor when someone like Lindbergh was honored with jets of water shooting from fireboats. A surge of strong emotion gathered in my chest, welled up as a sob that caught in my throat. If I had let go, let the eight days of tension reach my vocal chords, I'd have bawled like a baby. Instead, I choked it off with a wail that exploded as a shout of joy.

Dumbo came alongside, and a crewman at the side hatch tossed a rope toward me. It fell short of my grasp. The pilot had to keep the engine running; there are no brakes on seaplanes so it just kept gliding. He had to go around again to pick me up on the move. This time the crewman climbed out on the wing and dropped the line directly in front of me. I grasped it and hung on tight. Suddenly I was moving along at a five- to ten-knot clip with water spraying up around my bow. They pulled me in quickly; strong arms and hands reached down and lifted this dripping survivor into the plane's cabin, safe at last! The engines roared and we were on our take-off run immediately. The noise and my joy and amazement that I had actually met my goal of getting out in the water and being picked up had left me speechless; I could only grin. The New Zealand crewmen got me out of my wet duds and onto a bunk, wrapped a blanket around me and put a cup of hot tomato soup in my hand.

To an attending crewman I pointed out through the port the dark form of the distant island of New Britain and said, "I started out there in the hills of that island eight days ago."

He nodded as we exchanged grins. The loud engine noise drowned out any more words. I felt a deep desire to thank them all for their part in my rescue.

CHAPTER 4

IN SAFE HANDS

Sitting on a bunk in the Dumbo that rescued me from the sea, sipping the cup of hot tomato soup, I felt contentment slowly seep into my body. I was relieved. I knew the worry and concerns of my ordeal were over. I was in safe hands.

Stripped down to my skivvies, a blanket over my shoulders, almost dry; I was warming from the hot soup, the blanket, and the safe and secure feeling. A crew of seven or more surrounded me in various positions in the aircraft.

Outside, our fighter escort kept a watchful eye on the skies around us—except one—I noticed creeping in close on our port wing, really close. I moved nearer the window on that side, almost touching the glass. The pilot seemed to want to get close enough to see if he knew the guy they just pulled from the drink.

It might be one of my squadron comrades. I moved closer, trying to make out the details of his face. He peered intently through his canopy. Then he saw me. His mouth dropped open, eyes widened. He put his hand up to cover his face and turned his head away as if the sight were too horrible to bear. He and his plane turned away. Then I remembered what I saw in the mirror earlier in the day when I signaled the search planes. The face I saw was not a pretty sight, but of course that Corsair pilot out there just offered comic relief to this serious rescue. Or did he?

A little later, I looked out the port again, searching for landmarks. The northwest corner of Bougainville appeared just then, and as we neared the beach at Torakina I wondered if we would land there, if a reunion with my squadron was imminent. Would they be surprised that ol' JG came back to life?

I watched intently as the coastline slid by. I saw what looked like Princess Augusta Bay. We passed it by, and I realized we would not have a reunion at that time.

After about an hour, our fighter escort fell away. We were about to let down to our destination, a lagoon in the nearby Treasury Islands. We landed and taxied to a mooring. Soon a motor launch pulled alongside our port hatch and a medical officer entered. He looked me over, saw that I was ambulatory and got someone to hand me some clothes.

Dressed and warm, I climbed into the boat with the doctor and we shoved off in the direction of a destroyer-sized naval vessel anchored in deeper water of the lagoon. The name on the side read USS *Coos Bay. Hmm, that has a homey ring to it: the name of a small town on the coast of my home state of Oregon. I have cousins living there.*

We pulled up to the base of a steep ladder alongside the ship near the stern. The doctor told me to go first. I proceeded without hesitation until I reached the top, where I came to a halt. Memory nudged me with a warning that Navy protocol asks for something at this point: you must face aft and salute the National Ensign, then salute the officer of the deck and request permission to board. The reason I hesitated: another item of Navy etiquette: never salute unless covered, and I didn't have a hat. A gentle shove from my escort behind gave me permission to waive all protocol and etiquette just when I felt a swelling urge to display my patriotism.

The doc took the lead and we walked almost the full length of this vessel that I soon learned had the designation "Aircraft Tender", and into its tiny sickbay. Cleanliness and order flooded my senses. After more than a week mucking around in the jungle and sitting in water in a one-man raft on the big ocean, I almost felt like a contaminated alien in that ultra-white environment.

A corpsman tossed a towel, pajamas, razor, and toothbrush on my bunk. It appeared I was the only patient. The shower felt luxurious and I savored its cleansing qualities—unforgettable—such a pleasure to brush my teeth, not a chore at all. I dried off and put on the pajamas, enjoying the freshly laundered smell of clean towels and sheets.

The doc looked me over. I told him about the stuff swirling around inside my eye. He said it showed signs of an interocular foreign body. He noted that the torn nostril was almost mended, no need for stitches. I told him about the sulfa I had been taking. Superficial wounds on top of my head and my upper arm had almost completely healed. My good luck prevailed; wounds normally have a tough time healing in jungle environments. I guessed the seawater helped a lot.

The corpsman had me stand on the scales: I had lost twelve pounds. He helped me climb into my upper bunk to get ready for some sorely needed rest. They told me I was dehydrated, that I needed to drink a lot of juice, to help myself at the nearby fridge, but not too fast or I could get stomach cramps. I found it difficult to be cautious; my thirst had become insatiable. I was constantly at the fridge, ever warned by an alert corpsman.

I thought I would fall asleep once I slid between the covers, but it just didn't happen. *Is there such a thing as too much comfort?* I found the absolute heavenly feel of the soft bunk, cool, clean sheets and a *pillow*, almost intolerable. After six nights on the ground, on the rocks, in sand, and two nights in damp, drifting, rocking raft, this was just too cushy to bear. My body couldn't adjust; I couldn't stop squirming around, in the utter bliss of that newfound comfort.

Finally, a Navy Lieutenant arrived and introduced himself as an intelligence officer. He came to debrief me on my last mission. I described in detail the events on the day I fell from the sky and how I subsequently found my way into St. Georges Channel to be rescued. I told him the name and location of my squadron and he said he would send them a radio message that I had been saved and that my parents would also be notified of my new status from missing to wounded-in-action.

Oh, my parents, they think I'm missing. I hope they haven't had to worry about me too long... my poor Mom and Dad. And Roz. I hadn't put her name on the "next of kin" list, a bad oversight, because I didn't think anything would happen to me. Turned out the intelligence officer subsequently released the information to the press, and she heard about it on a public broadcast, that I was safe. She hadn't even known I was missing.

At about dusk they brought in a meal for me. I ate half of it and I was full. I thought that my stomach must have shrunk or something.

Shortly after, another patient was brought in. They had to treat him for injuries and set his broken leg. Rescuers found him clinging to a rock near Buna Point, the sole survivor of a mission trying out a new five-inch cannon in the nose of their B-25.

The case of extreme comfort I was bearing, exacerbated by the ado over a cast going on the leg of my fellow patient, along with images of my survival journey and worries about my injured eye, all raced around inside my head and kept me awake most of the night. I wondered about my future as an aviator. I wouldn't be returning to my squadron. I knew it in my heart.

The next morning the medics told me I would board a SCAT plane that day, bound for Mobile Hospital Eight on Guadalcanal. I had left that island just nineteen days ago on a SCAT plane headed for the combat zone after a good night's sleep at the Hotel De Gink. It wasn't as luxurious as it sounded, just a Quonset hut with rows of cots and clean blankets. Now I would be going to a different kind of hotel: Mob 8, they called it. *Just another Quonset hut with rows of cots?*

SCAT was short for South Pacific Air Transport Command. They flew all over that part of the Pacific carrying everything you could imagine: cargo, passengers, aviation fuel, wounded men, airplane parts, weapons, food, and medicine. They, among others, helped win the Battle of Guadalcanal, our then tenuous hold on territory dangerously close to our Australian and New Zealand allies.

Now I would return to this historical island on a military DC3, not as a passenger but as a hospital patient, on a stretcher. Circumstances had dealt a whole new hand for me. I felt strangely ambivalent; lucky to be alive and well cared for, but unable to shake a sense of malaise that gripped me.

Only twenty-four hours before, I had been floating in enemy seas, hopeful of an imminent rescue, alert and deep in a survival mode. I had been on my own for eight days. Now, suddenly, I had a whole new purpose, a passive, but worrisome one: to heal and overcome possible infection in my damaged eye, with the help of the Navy Medical Corp. Shifting purpose wouldn't be easy after the active life of a fighter pilot who has just survived a life-threatening experience. I can't shake a growing, empty feeling that I am suddenly out of work.

Then I got a little task to perform. My rescuers from yesterday had requested my appearance on deck at about midship for an informal ceremony. As I approached, I saw four or more of them clustered around a small, yellow, inflated life raft. It had some of my belongings in it, I saw as I arrived on the scene. They were respectfully returning my gear to me, including all of the survival equipment. All I really wanted was my personal stuff, my wallet, the knife my dad made for me, my wristwatch. I felt so grateful to those guys for their part in my return to safety I would have given them anything, even gear I didn't own. Those Kiwis could have it all, even if it did belong to Uncle Sam. I thanked them all profusely, then sauntered off with my prized possessions. Back in the sickbay I looked through my dried-out wallet. In it still were cards and pictures and one hundred dollars in damp bills. The pictures of Roz had survived. *I must write her and tell her I'm okay.*

After looking at the pictures long and hard I stuffed the wallet in my pocket, slipped the belt with knife sheath through trouser loops, slid the waterlogged watch into my pocket; I was all packed for the flight. I loved traveling light.

They took me to the airfield on Stirling Island where I embarked on the familiar SCAT transport, now rigged with stretcher racks specifically for carrying hospital patients. We were soon underway, bound for Guadalcanal and Mob 8.

Navigation was easy there; we just aimed down "The Slot"; that well known, highly visible waterway between the larger islands in the Solomon group. We passed between Choiseul and Vella Lavella, between New Georgia and Santa Isabel. Our course took us north of the Russells and south of Savo Island on a heading of 115 magnetic degrees right into Henderson Field on Guadalcanal.

Fellow patients and I were taken by ambulance from the airstrip, along palm shaded roads between Quonset huts, past Hotel De Gink, the Quonset TOQ with the ritzy name, and into a compound comprised of a cluster of more Quonset huts, my new temporary duty station, Mob 8. I was assigned a cot in the middle of a neat row of beds, mostly filled with wounded and ailing Marine and Navy officers.

Later in the afternoon I had visitors. Two familiar faces approached my bed. They were Doc Donnely, our squadron Flight Surgeon and Major Kempson, the Exec. I was so glad to see them that I started to hop out of bed and they tried to stop me.

"No, no, stay in bed. You don't have to get up," they said, and I settled back.

They couldn't stay long, had to get back before dark. Doc looked me over.

"They did a great job on your nose," he said.

"It healed itself," I replied, "no stitches."

I told him about the special first aid kit he had made for each of us in the shoulder harness, that it ripped off when I bailed out, how I managed to make-do with the stuff in the jungle kit. He could see my wounds were healing well. I told him about the interocular foreign body and my impaired vision.

"All the guys were in the 'ready shack' when the call came in about your rescue and a great cheer went up… we had a long celebration that night," Major Kempson said as he handed over a sheath of official looking papers and my log book, "Looks like you'll be heading home." I detected a hint of envy in his voice.

They had to leave. "Say 'hi' to all the guys. I'm going to miss all of you." We shook hands again. I watched them leave all the way to the door. They had made a long flight just to get my papers to me. I felt rewarded by this all-too-brief visit from comrades.

I thought about all of the guys in the squadron flying together for more than six months. We had a good record with minor accidents, a couple of crackups but no fatalities. Not like other squadrons who had lost pilots in training accidents. We weren't used to losing comrades. Now we had, in our first three missions, three guys who didn't return: Bob Thompson, Joe Hennaberry, and Zeke Miller. It raised the spirits up there on Bougaineville when a guy like me showed them they could come back.

As I thought about those first combat missions, running back through them in my mind, I caught sight of a corner of my logbook peeking out from under my pay records there on the bedside table. I picked it up. It had a record of all the flights I had ever made. I turned to the last page. All of the latest entries were in red ink, signifying combat flights. The last entry caught my eye… it said, *"missing in action."* Before that it listed the date: *February 10,* then the type machine: *F4U-1,* the bureau number: *02566,* the duration of flight was blank because they didn't know that; character of the flight: *combat,* name of pilot: *Morris, J.G.* and finally, under remarks, the name of that mission: *Vunakanau strike. Hmm, my fourth mission. I had more experience as a survivor than in combat flying.* The days that led up to that last mission were chronicled briefly there in the log. I looked at the first entry, dated 1 February.

FEBRUARY 1944

Date	Type of Machine	Number of Machine	Duration of Flight	Character of Flight	Pilot	PASSENGERS	REMARKS
1	DC-3		:40	N		J.G. Morris	
4	F4U-1	55722	3.0	"	Morris J.G.		Tobera Strike
4	F4U-1	55716	2.5	"	"		Task Group cover
6	F4U-1	55720	3.0	"	"		Lakunai Strike
10	"	02566	"	"	"		Vunakanau Strike
10					Missing in action		

	This Month	Total Time
	85 4.0	12.0
	531.0 86.0	352.0
	532.5 88.0	369.5

Total time to date: 85

I CERTIFY THAT THE FOREGOING FLIGHT RECORD IS CORRECT.

_____Feb. 29, '44_____ (DATE)

J.G. Morris (Sig)

THE FOREGOING FLIGHT RECORD IS APPROVED.

BY ORDER OF THE COMMANDING OFFICER.

On our flight into the combat zone on Bougainville, as a passenger on a SCAT DC3, I could see through the window as we circled the Torokina part of Princess Augusta Bay, small puffs of white smoke interspersed with larger black eruptions of artillery, mortar, and grenade fire. We would be landing on one of three strips built by The Sea Bees (the Navy's construction Battalion), on a perimeter conquered by Marine invaders and now held by the Army. The Japanese still occupied the bulk of that large island with a rumbling, smoking volcano at its highest point.

On the ensuing ten days we heard the sound effects: sharp cracks of rifle shots, banging grenades, booming artillery fire in the distance, and the deep rumbling of Mother Earth far below, while we went about our business preparing for missions, distant targets on other islands. After dark we could expect a visit from Washing Machine Charlie, a Japanese night bomber, who sometimes dropped a bomb close enough to interrupt our sleep. In our safe perimeter we could do almost anything. I got my exercise body surfing in the warm, tropical waters of its bay, relaxing afterward on the black, volcanic sands of its beach.

After a brief moment of reverie, through half-closed eyes I saw the logbook had fallen open to another page, the page before, the one showing the last entry in January, the four hour flight from Espiritu Santo to Guadalcanal in an *F4U-1,* bureau number: *02566,* the plane I landed on Henderson Field. *I remember that plane and the landing.* It had made me nervous. Something was not right; I had just flown it six hundred miles over water and the engine made little popping sounds every time I lowered speed. As I entered the traffic pattern at Henderson it really popped when I throttled back; it backfired. I was relieved to get it down on the mat, a steel mat that followed the contours of the land like a miniature roller coaster. I taxied to the flight line and turned the

January 1944

Date	Type of Machine	Number of Machine	Duration of Flight	Character of Flight	Pilot	PASSENGERS	REMARKS
11	F4U-1	02772	1.5	E	Morrie J. G.		test flight
11	F4U-1	55598	1.5	R	"		
12	F4U-1	55429	1.5	ZB	"		ZB - Homing
13	F4U-1	55901	2.0	ZB	"		"
14	F4U-1	55901	1.5	F	"		"
19	F4U-1		1.0	F	"		STRAFE
21	F4U-1	56076	1.5	F	"		WATER INJECTION TEST
22	F4U-1	55932	1.0	Y	"		PRE-DAWN TAKE-OFFS
22	F4U-1	55933	1.5	F	"		STRAFING
23	F4U-1	55934	1.5	F	"		"
26	F4U-1	55938	1.0	Y	"		PRE-DUSK TAKE-OFFS
31	F4U-1	02560	4.0	"	"		FERRY FROM ESPIRITO SANTOS TO GUADALCANAL

	Pilot	Passenger	Total Flying Time
TOTAL THIS MONTH	19.5	—	19.5
BROUGHT FORWARD	511.5	26.0	537.5
TOTAL TO DATE	531.0	26.0	557.0

I CERTIFY THAT THE FOREGOING FLIGHT RECORD IS CORRECT.

7/1/1 (DATE)

J. B. Morrie (Sig)

THE FOREGOING FLIGHT RECORD IS APPROVED.

FLIGHT OFF.

BY ORDER OF THE COMMANDING OFFICER

snorting beast over to a plane captain at the fighter pool. This plane was not one of our squadron planes. It had been at the Turtle Bay Fighter Strip in the New Hebrides for overhaul. It needed more work before its return to combat. Happy to be out of its birdcage cockpit, I wrote it up on the flight report and caught a ride to the Hotel De Gink.

I looked down again at the bureau number. This was one of the earlier Corsairs out here. Its number started with a zero. I turned the page to see what the numbers were on our squadron planes. They started with the number five: brand new planes that came with us from the States on the USS *Barnes*, the baby flattop. Both decks were loaded with twenty of the latest models, equipped with water injection. They were now the hottest planes in the Pacific Theater. Our squadron was the first to fly them.

I read on. On my first three missions I flew one of our new planes. *But, look at this. The plane I flew on my last mission starts with zero—not five. It was a really old plane... a replacement for one of the three we had lost already. Newer planes have bubble canopies; old ones have what we call bird cages.*

Then my gaze fixed on that last entry; the bureau number, 02566, had a familiar ring. *I wonder why? It was old... no wonder I had a hard time keeping up with my section leader... but what else?*

I turned back to the previous page, my focus on the over-water ferry hop, the bureau number, 02566. *My God, that plane that sputtered and popped may have been trying to tell me something!*

I continued to stare at that number. I read it over and over, 02566. It had a familiar sound, like another number I knew real well... like... like... *Hey I know! My Marine Corps serial number: 025665. Why, that plane almost had my number.*

I heard later that as a result of my reporting about the difficulty keeping up while flying an older plane, they made the decision never to assign an old plane to a wingman again. Wingmen are followers. *But they have to keep up!*

JG MORRIS

N. AER. 4111

AVIATORS FLIGHT LOG BOOK

2nd Lt., J.G. Morris,
USMCR, 025665

NAME
RANK
BRANCH OF SERVICE

A couple of days later I noticed corpsmen scurrying around putting extra attention to dusting, swabbing decks, straightening sheets and blankets. I asked one of them if we expected an inspection because of Washington's Birthday.

He said, "You haven't heard? Nurses. Their very first visit to Guadalcanal. They'll be stationed here."

So, nurses would be part of an inspection. I was sure it would be a two-way job. Some of the guys hadn't seen a woman in months. I started thinking about my appearance, wondering if I could do anything to improve it, hoping they didn't expect too much from us patients. I decided to check it out anyway, in a mirror in the head. My distorted vision didn't help. I hadn't shaved in a couple days; could do that. The part of my right eye that one expected to be white was still crimson. The bags under both eyes had a purplish haze. The skin on my face was mottled from sunburn that had blistered in places, peeled, then burned again. The scar on my broken proboscis stood out like a sore thumb... not much I could do to get ready for that inspection. Back in my bed I turned to reading, trying to forget my appearance.

The corpsman came back to inquire about my wounds, "Do you have any wounds that are obviously from gunshot? We are checking for your eligibility to receive the Purple Heart."

I showed him the place on my upper arm where I had squeezed out a pea sized metal piece. He entered the least serious of my wounds on the form and departed.

Another corpsman came through spraying DDT with an aerosol can and we were supposedly ready for the big inspection, no mosquitoes allowed.

I could see the entourage as it entered the building way down at the end: nurses in seersucker uniforms with thin navy stripes, a senior medical officer in khakis leading the way. They stopped at each bed, greeted its occupant cheerily, then moved on to the next. They were getting closer. I leafed through my logbook, trying to show an air of casual nonchalance to cover real feelings of self-consciousness.

Suddenly, women were standing at the foot of my bed with a Navy Captain who stirred up my angst with, "Lieutenant Morris? Hmm, you don't look very pretty but, here's something pretty for you."

He held something out toward me. I started to get up. He said, "No, no, you don't have to get up." But I was already up, eagerly looking at a medal hanging from a purple ribbon draped over his fore finger. It was the Purple Heart with the profile of the man who invoked the medal during the Revolutionary War, George Washington, whose birthday we were celebrating. It was another patriotic moment that called for a salute but of course, I was not "covered" again so I simply stood at attention as I reached for the gift. The Captain laid it in my hand and I thanked him.

The group moved on. I watched the movement with quiet reverence, then turned my attention to my new gift. It was a gold-rimmed heart enclosing a deep purple heart in which nestled a handsome golden profile of our first President. At the top center, a gold-rimmed white shield enclosed three red stars and two red stripes. The red signified the flow of blood. Many of these medals were awarded posthumously and sent to parents and loved ones. My parents and I were lucky.

I wrote to them the same day and told them I was okay, just in case that wire didn't get through and they were inadvertently told I was missing. An American Red Cross person supplied me with pen and official stationery with a red cross at the top and I wrote:

Feb. 22, 1944

Dear Mom and Dad,
Haven't been in a position to write during last few days. Now finding myself resting and receiving a little eye treatment at Navy Hospital. I really feel wonderful, but if they insist I'll take it easy for awhile.

Please don't be alarmed by any report of my "missing" state of affairs. I'll tell you about it all when I get home...

I sent a "V" mail letter to Roz the same day, telling her the same story about being in for eye treatment and not to worry.

The doctors checked my eye as best they could with a hand ophthalmoscope and confirmed that there was an internal injury and possible foreign bodies.

They wrote up orders to send me to another, more advanced mobile hospital in Noumea, New Caledonia, about a thousand miles south. I would be taking another of those all-purpose SCAT planes way past the New Hebrides right into the center of an imaginary triangle connecting Australia, New Zealand, and Guadalcanal.

Once again I sat in a plane leaving Henderson Field, this time heading south, away from the combat zone. As we rose up over the ridge to the southeast I realized we were passing over expensive real estate: land bought at the price of many lives, spilt blood, suffering, and hunger. That was no ordinary ridge. That was Bloody Ridge, a place where the Marine Raiders bravely kept the Japanese from retaking the airstrip, a year and a half before.

On the long seven hundred mile flight to Efate in the New Hebrides I napped and dreamt about the many flights our squadron had made in the air around the Turtle Bay fighter strip. I had my logbook out, scanning the entries of January. I saw test flights, water injection, formation, strafing, and pre-dusk flights when we took off in late daylight and returned after dark to make night landings.

In early January 1944, VMF 218 had disembarked at Turtle Bay fighter strip and taken possession of twenty sleek new F4U-1 Corsairs. In a few days we had them in the air, breaking them in (not up), and gaining more combat training in preparation for the "real thing" a-coming. We strafed rocks along the shore, intercepted pseudo bogies, chased tails (a form of follow the leader), and played chicken with sudden tropical squalls at the end of runways.

All that high-powered activity suggested, of course, the possibility (probability?) of more misadventures. Until now our only pre-combat casualties had been minor injuries e.g., a gunsight imprint on McCabe's forehead (he got from a landing in the boonies back in the states).

On January 22, we joined-up in the dark after a pre-dawn take off. At 10,000 feet directly over the field, we heard our leader's voice in our headsets telling us to join-up. We looked around and it seemed as if everyone was there, a lot of planes in close formation.

Someone called back, "Yokel green leader, what is your position? Over."

"This is Yokel green leader. I'm directly over the field at ten angels! Where the hell are you guys? Over."

"We're directly over the field at ten angels all joined-up looking for you, sir. Over."

This was a mock combat situation. We were supposed to be "orbiting," waiting to be "vectored" toward "bogies" by the radar command base. Darkness had faded and our leader awaited our join-up. Then we got the alert from the base that bogies (pseudo enemy planes) were already making a run on the field. I was on Spindler's wing. The two of us dove straight down for the deck. We could see that it was too late to intercept the bombers before they reached the runway. We aimed for the other end of the runway. I glanced down at the air speed indicator, to see the needle at 550 knots. I had never flown this fast. We leveled off just above the field and streaked between the first and second bogies. We were going so fast our planes jumped when we tried to bank (the air does strange things when it piles up in the front of wings). We pulled back on our sticks to slow our planes and get them under control, but climbed so fast we flew out of firing range in an instant.

After landing we felt depressed, not so much because of our tardiness in intercepting the bogies but because we couldn't figure out how our leader, with his wingman, knew exactly where *he* was, while eighteen of us were all lost together.

A few days later, a group went out stunting over the island in different formations. This was the day Enders "joined-up" by accident. They went into a "tail chasing" formation. Diving and climbing single-file resulted in radical changes in airspeed, making it hard to maintain exactly the desired distances between planes. The pilot tailing Enders lost sight of him temporarily. Then suddenly he saw the other plane's tail coming down on his propeller and in a flash the tail flew off and Enders found himself spinning toward the ground.

He hit the silk all right and landed in a clearing a number of miles north of the air base. After handing over many gifts of appreciation (jungle kit, parachute, rubber raft) to demanding natives, he was led safely to the Frenchman's plantation on the coast. The natives would have had Enders' rip cord too, but he clenched it too tightly in his fist.

A few years later I found out James Michener made up tales of the South Pacific on this same island, a short distance from the Turtle Bay strip. One of his stories involved a French plantation owner, too.

Later, a small formation of VMF 218 Corsairs took a heading toward "the slot" on its way to Guadalcanal, away from the combat zone. No need to fly close; loose formations took the strain out of long hops.

Previous word from Intelligence warned that some small islands just beyond the Treasuries, bypassed by our ground forces, still had a few surviving Japanese and they reportedly had antiaircraft guns. Now this flight had just passed the Treasuries, approaching one of those little islands. Their course, if held, would go directly over it.

Over the air came uneasy comments, "Hey there's that island we're supposed to go around. Shouldn't we kind of slide off to the starboard?"

No word from the flight leader.

Someone surmised, "He's not tuned in."

They kept calling and got no answer.

Another suggested, "Let's just loosen up the formation. Kind of slide over to the starboard. Waaay over."

"But... but... what about him?"

"Aw, let 'im go. Why should we spoil *his* fun?"

The first leg of the long flight was almost over when our plane let down and we passed over clusters of small islands in the New Hebrides. We bypassed Espiritu Santo and made a quick pit stop at Efate, then moved on toward New Caledonia.

After another long flight over water we landed at Tontuta airstrip in the middle of the long, narrow island that held our new temporary station on its southerly tip at Noumea town. Surface transportation awaited us in the form of olive drab ambulances with red crosses on white circles. All ambulatory patients climbed into the ones with bench seats lining both sides. Patients on stretchers were taken into real ambulances.

Noumea sported one of the biggest and busiest harbors in the South Pacific: ships of many nationalities anchored there in that French Possession. Large seaplanes landed and took off at the old Clipper terminal. Small shore boats dashed about from moored ships to docks, merchant ships unloaded supplies at the wharfs, dry-docks berthed damaged warships under repair. The town was the closest thing to a city I'd been in since I left San Diego, but we wouldn't really be in town; I could see it on a hill: MOB 11 stood at its vantage point above.

The hospital buildings were not Quonset huts there, but long, low, single-story, pole structures with thatched straw roofs, a touch of ethnic warmth. Inside, waiting to be admitted, I found a number of beds clustered close to the nurses' station. In one bed I saw a young man sitting in its middle, holding his bandaged leg.

He told his story while I waited by his bed: he was a PT Boat officer, stationed in Empress Augusta Bay close to where my squadron was based. A mooring cable snapped, whipped across the deck and ripped off his leg below the knee. He seemed to have taken this traumatic loss with a strange sense of humor.

"I don't know what they did with my leg," he said, "but the Skipper's dog had been howling every night for a week but stopped the night of the accident. He must have got what he was howling for."

"A tough old nurse," he went on, confidentially, "came on duty my first day here. I was sitting here like this with my leg tucked under me. She asked me what my complaint was and I told her 'No complaints.' She saw me holding my legs and asked me what was wrong with them. I told her, 'Not much, one's just a little shorter than the other'."

He introduced me to the older Warrant Officer in the next bed who, I soon learned, had a glass eye and false teeth. The WO flew with Richard Byrd, one of my earlier heroes, at Antarctica and lost his eye in a flying accident before he went to the South Pole. I didn't ask about his teeth.

I had a special interest in how flying was affected by loss of vision in one eye. I asked how he did it.

"You can duplicate depth perception by moving your head side to side," he told me, "but landing on snow is another matter. When the sky is white, with no visible horizon, flyers with two good eyes have no advantage in the snow. The only way is with a power let-down. Just feel your way down and chop the throttle when you touch down."

I remembered making water landings at night the same way. Maybe if I didn't get all my vision back, they'd still let me fly. The old guy gave me hope.

On my way out, Bill, the PT guy, had more to tell me, quietly, "It's a kick watching the old guy get ready for bed. He takes out his false teeth and brushes them, puts 'em in a glass by his bed, takes out his eye and cleans it, puts it in a small box and he's ready for sleep."

All the beds were taken in that long ward, so a corpsman took another guy and me to the adjoining ward, one long empty room. Looked like we'd be roommates. We took adjcining beds in about the middle of that quiet space, stowed our gear in bedside cabinets, sat on our beds facing each other, and exchanged stories.

This was different than being part of a squadron. In a hospital your companions have a wide array of complaints, reasons for being hospitalized: some had combat wounds, wounds from accidents, diseases such as malaria, dengue fever, jungle rot. Some had no visible ailment but they suffered from varying degrees of mental and physical fatigue. This was the case with my new friend facing me there in that empty ward.

He wore the khaki uniform of a Navy Lieutenant: short sleeved shirt open at a collar bearing single silver bars, and gold wings on his chest—a Naval aviator. We had something in common.

He took out a pack of cigarettes and offered me one. I took it, even though I didn't smoke, because it looked like there wouldn't be much to do until supper. Thus began an addiction I would have for another nineteen years, as an antidote to boredom. And that was his ailment. Seriously. He suffered from an acute case of boredom—well, not just boredom.

He had been out there flying sub patrol in a dive-bomber for almost two years, protecting the vast amounts of shipping that plied the waters of the Coral Sea off the coast of Australia. It was boring, but something else too: a mixture of apprehension, fear, anxiety, loneliness. He flew alone—no companion planes alongside in formation—the loud roar of that single, bleating radial engine droning on day after day. Even worse, the droning might suddenly stop mid-flight, calling for a mid-ocean, wheels-up belly landing. His diagnosis was something like "nervous exhaustion" or "fatigue." He just needed a break before he went nuts, like some of the guys who were isolated from the rest of us in a secret place usually designated "Ward Eight." Luckily, they caught my new friend's symptoms in time. He inhaled deeply on his cigarette with what appeared to be a profound sense of relief. I puffed on mine as I recounted my own recent traumatic experience.

By the time afternoon "story hour" had played out, the dinner hour arrived, and we made our way down a slope to a nearby hut from which drifted the familiar smells of the Officers Mess. My appetite had grown to gargantuan proportions and I found myself returning for refills, doubling up on dessert: two slices of apple pie piled high with vanilla ice cream. I could fix my weight loss situation easily with the close proximity to the mess. And that was about the extent of my activities for the next three weeks: puffing on borrowed cigarettes, waiting for wounds to heal, waiting for the next meal, and ultimately for travel orders.

CHAPTER 5

ACROSS THE PACIFIC

When the orders came, I found I had options: I could pick any mode of transport available—to the Naval Hospital of my choice—in the States. I knew at once I wanted to be as close to Roz as I could and that meant San Diego and the hospital there, about a hundred miles south of L.A.

I proceeded to the nearest air base, which happened to be in Noumea Harbor, a seaplane base, to seek out a flight toward Southern California. They told me a flight would leave on the morning of 15 March and reserved a space for me. Destination: Pearl Harbor in Hawaii, many, many miles closer to my loved ones. In a matter of days I'd be with Roz again. *What would it be like? Things were different now. How different would it be?*

Two days later my flight was under way aboard a Navy Martin Mariner, a two engine, twin finned flying boat carrying eight passengers. We had a Pan American crew, veterans of the prewar Clipper fleet, which included a steward who cooked and served meals in a small galley. Our long flight across the Pacific would be interrupted by four stops, the first of which was at Luganville on

Espiritu Santo.

When I learned this stop would be about an hour and a half, visions of my dress uniforms in a suitcase stowed in a warehouse at Turtle Bay Fighter strip came to mind. *I can do it,* I thought, *Hitch a ride, with any luck, get there in twenty, thirty minutes... check out my suitcase... and return. It is a little chancy but I can save myself a lot in time and money. I won't be able to take the footlocker... too heavily laden with about ten fifths of Old Forester. Leave it for the guys.*

As we tied up to the fueling dock in the bay at Luganville, I had made my decision. In minutes I stood with my thumb out, looking down a well-traveled road. My good luck prevailed and I was soon on my way to Turtle Bay—on a uniform mission.

Upon arrival there I immediately sought out the storage area where personal gear was kept, sometimes they called it "personal effects." I thought about the poignancy of that when a guy showed me where my gear was stowed and I saw my name stenciled in black letters on the side of the light green carryall suitcase. I remembered a dramatic moment a scant few weeks before when I stumbled over a dark green duffle bag as I rounded the corner of one of the huts in that very camp. The name stenciled in black stood out vividly: MAJOR GREGORY BOYINGTON.

They were his personal effects. He had been shot down on 3 January in St. George's Channel and captured by a Japanese submarine. Boyington was already a legend out there; we all knew about his exploits as commander of VMF 214, the famous Black Sheep squadron, and were saddened by the loss.

It all happened so fast: I got my ride to Turtle Bay, picked up my suitcase, then I had a little time to spare. *I'll make a quick stop at the Intelligence hut; maybe get a chance to thank the survival instructor for the valuable tips I received there; especially the time spent working with the intricacies of the signal mirror.*

From the top of the stairs, I could see into the large room inside. There was a gathering of pilots with the survival instructor standing at the front, lecturing. I entered and waited in a small foyer, reluctant to disturb the class. He might be almost finished. I couldn't hear what he was saying but I did catch a word or phrase once in a while. I heard, "St. George's Channel" and "Cape Gazelle" and some other familiar ones.

I looked at the clock on the wall and saw that I had to move on if I expected to make it back before departure time. I still had to hitch a ride.

I turned to leave and as I started down the stairs, the meeting broke up. One of the guys, apparently in a hurry, rushed past me. I recognized him from flight training. We exchanged greetings.

I asked him, "What was the lecture about?"

"Oh, it was good news. A fighter pilot, shot down over the Gazelle Peninsula near Rabaul, reached the coast, launched his raft and got to a place for a safe pickup by Dumbo."

"That is good news. I'm glad he got away."

And now he has to hitch a ride in a jeep to another seaplane, I said to myself as I hustled off toward the outbound road. I would have liked to tell him I was the rescuee but I felt the press of time.

Back aboard our seaplane on the old Clipper route I got acquainted with fellow passengers and found I was a candidate for the Short Snorters Club. To become a member you had to be on a long, over water flight with at least one other person who was already a Short Snorter. All members got to sign your pass—which is a one-dollar bill—and you got to sign theirs. Some of the passengers had long Short Snorters made up of two or three bills taped together, full of signatures, and they had to keep them in their wallets. If you ever claimed to be a Short Snorter and couldn't produce proof, you had to buy drinks for all other members within earshot.

The inscription around the border of my new pass told where this ritual took place: the latitude, longitude, and altitude aboard a PBM. It had signatures of crewmembers and passengers including one woman, a TWA flight attendant.

Our next stop was in the Ellice Island Group at Funa Futi where we took on fuel. We didn't even disembark, just sat there rockin' at a dock while they pumped it to us. In minutes we were back in the air, winging our way across the International Date Line into yesterday. On a map in the galley I could see that we would cross the equator sometime during the night. Word was we would make a rest stop at Kanton Island, just below the equator. As I casually scanned the map, something caught my eye, a familiar name: Howland Island. Radio news broadcasts in the late thirties kept repeating that name. It was where Amelia Earhart got lost. She was trying to reach Howland Island and now we'd go right past it on our way to Pearl Harbor.

On our stop at Kanton in the middle of the night we disembarked to a dock and walked into the Pan Am terminal where they had cots with blankets for everyone, including the crew. All was quiet. We were tired and enjoyed a respite from the noisy engines.

As I lay back and closed my eyes, thoughts of Earhart came to me. We were so close to her final destination, maybe closer than we knew. And I thought about those early flyers, men and women who took risks that later made air travel more feasible for all of us. Radio sets were rare in 1927 and we had to go to Uncle Will's house to hear about Lindy's solo conquest of the Atlantic. I heard the radio broadcast that announced the fatal last flight of Wiley Post in Alaska with Will Rogers aboard on a daring international flight. Our own little military airfield at Fort Vancouver, Pearson Field, was the destination of exciting early flights.

When I was about eleven, we neighbor kids were on our way to school one spring morning when we heard something unusual in the air. We had heard the sounds of aircraft overhead before, but nothing like this. We looked up and caught sight of a large group of planes flying together in even formation, in vees, like a flock of geese. They were small planes with two wings and one engine, but together they made a lot of noise and a sight to remember.

They seemed to be letting down toward Pearson Field. They were landing at our town and that became a magnet for every kid around. We migrated south like birds, past the street our school was on. A field trip just happened: our principal was a boy that day, too, and declared a school holiday. We all gathered on Pearson Field and inspected the Boeing pursuit planes that flew from the factory in Seattle and the dashing fighter pilots, with helmets and goggles and white silk scarves, who flew them. A day to remember!

That same little field became the unexpected destination of a famous international flight in 1937 when a Russian multi-engine plane made the first great circle flight over the North Pole. They meant to land at a more prestigious site like San Francisco but a low fuel supply had them looking for anything available. Portland's airfield was socked in, giving Vancouver its shot at international fame.

For many years a famous person secretly flew into that field on a regular basis and took a cab to the nearby village of Ridgefield, Washington to visit a housebound relative who lived there.

That same person had secretly visited our squadron when we were still in training at the Marine Corps Air Station in Mojave, California. Looking out the window of our ready room toward the field, I caught a glimpse of a Corsair landing. It had my immediate attention because it was the first time I had seen one with a belly tank (a reserve fuel tank carried on the bottom between landing gear). So I kept watching it as it landed and taxied in our direction. It pulled into our ramp and a crewman guided it into the flight line. Curious, I kept watching as the pilot prepared to get out; it looked like he wore a coat and tie. He reached back into the cockpit, lifted a briefcase, and put on a hat and climbed down. As he strolled toward our building, it slowly dawned on me who he was, this tall slim aviator with sandy hair. He had come unannounced and I watched him enter the building. I stood in the ready room doorway as he walked right past me, and traversed the length of the hall to the Skipper's office. Not a word was ever said about this visit; officially it never happened but we did admire the smooth take-off that came a little later by a phantom Corsair with a belly tank doing a "Lindy Hop."

Then I thought about stumbling over Boyington's "personal effects" back at Turtle Bay, about his part in aviation legends. They called him "Pappy." He had flown with Chennault in the renowned Flying Tiger squadron as a volunteer in China before Pearl Harbor: a colorful swashbuckler, skillful pilot with a reputation as an independent thinker. They gave him command of a collection of young pilots who happened to be adrift there at Turtle Bay, unattached to any squadron, just waiting for this moment to make a difference in aerial combat in the upper Solomons and Bismarcks at the end of 1943. "Black Sheep" was their name and "Pappy" brought them fame. They called him that because of his age. He was 28, 29 and they a mere 19 to 21. The squadron had a "turkey shoot" around Rabaul and Bougainville and earned a reputation as a highly productive outfit. In other words they knocked down a lot of "bogies."

Later, in Boyington's autobiography, I found some interesting coincidences. When he was shot down on 3 January in St. George's Channel, a Japanese submarine crew picked him up very close to the spot a sub went past me in the middle of the night of 17 February. He was held prisoner in Rabaul for about six weeks. They knew who he was; he had a reputation with them, too. They withheld treatment of his wounds for ten days (he had an acute scalp laceration) as a form of punishment just for being Boyington.

Coincidentally, on the night of 16 February we both left the Island of New Britain, he as a passenger in a Japanese Betty bound for Truk, I as a refugee in a one-man life raft bound for safety.

The Betty landed at Truk at the same time the U.S. Navy launched an aerial attack on its airfield. When it landed, it screeched to a halt, everyone scrambled out into a ditch alongside the runway, turned, and saw the plane blow up behind them.

Boyington survived in a prison camp in Yokohama and was liberated at the war's end. According to his account, he befriended an elderly Japanese woman who worked in the kitchen and smuggled extra food to him under her apron. He actually put on pounds, while a prisoner of war.

Back home, he was rewarded with a promotion to Colonel, the Navy Cross, the Congressional Medal of Honor, and a reunion with his Black Sheep.

After having all those aviation legends buzzing around in my head, I wasn't quite ready for an early predawn take-off from Kanton. But the smell of bacon drifting in from our airborne galley soon brought me back to the present. We had a casual breakfast in shifts (the small dining cabin seated four at a time) as we headed for our next stop at Palmyra, nearly a thousand miles to the northeast. This leg was uneventful except for a spectacular sunrise under the starboard pontoon. Morning sun in my eyes brought drowsiness, so I dozed the last few miles away until we landed in a romantic-looking lagoon about mid-morning. After a brief refueling stop we were soon on our way again on the last leg of our flight: Pearl Harbor.

A little more than two years before, on a Sunday morning, Japanese pilots had winged their way toward this same place, not as a destination but as a target. They approached with hostile intent: to let loose a barrage that would destroy the bulk of the United States Fleet in Pearl Harbor that brought us into war and changed most of our lives.

CHAPTER 6

PEARL HARBOR AND BEYOND

I thought about how those events changed us, how I was transformed from college student to Naval Aviator. Two weeks after "Pearl Harbor," I was in Portland, at home for the holidays in the middle of my second year at the University of Oregon. I had registered for the draft six months earlier and reported in again to check my status, to get an idea how much time I had left. They had me classified 1A and my call-up was imminent. I would be drafted shortly after Christmas. The thought of Army combat did not appeal to me, and I headed immediately for the Navy Recruiting Office, where I took the physical and filled out all the papers. But I lacked one important piece of paper: my birth certificate. The officials there told me I could have an affidavit signed by a parent and notarized, and it would take care of the missing certificate. They gave me the necessary blanks and I hurried off on an urgent quest. I thought of going to the foundry where Dad worked and have him sign right there, but I decided to wait until he came home that evening.

I arrived home to find my mother in a heightened state of

excitement. She was so glad to see me and to know my enlistment was delayed because she had heard about a new program in Navy recruitment, aimed at candidates for flight training. College students with one and a half years of college could now apply and if accepted, would be given a draft deferment while they finished their second year. Perfect! This was just the thing for me, as long as I passed the physical and took care of business before the draft caught me.

At the recruiting office I learned more about the program known as V5. Candidates were even offered a bonus of $500 for each year of active duty they served. In June, we would report for preliminary ground training at an institution called Preflight School, as Naval Aviation Cadets. I would have to go to The Naval Air Station at Sand Point in Seattle for a special rigorous flight physical.

Once I signed up as a candidate, they took care of notifying my draft board of my new (temporary) status, issued me transportation and billeting tickets, and I was soon on a train to Seattle.

At Sand Point I entered a Naval establishment for the first time and found that they had only seaplanes that took off and landed on the waters of Lake Washington. The medical center was equipped with state-of-the-art testing devices. They tested for visual acuity, depth perception, peripheral vision, and color blindness. They had a special spinning chair you sat in and were twirled until you were dizzy, then checked to see how quickly you recovered your equilibrium.

When the Doctor looked at my teeth, he said, "Uh-oh. A bridge. This disqualifies you for flight training, but the ban has been lifted and we are now authorized to issue waivers. Also, your blood pressure is too high but we'll wait a bit and check it again."

He had me sit and rest awhile, then took my pressure but it was still above qualifying level. The Doc did everything he could to reassure me, then took it again—still too high and time had run out. The workday was over, time to go home. The Doc, prepared to give me another chance, knew my physical condition was excellent and that I was just nervous. He said, "Go back to your hotel, get a good night's sleep, take a brisk walk in the morning, and come back for another check. I'm sure that will do the trick." He was right, a good guy who helped me over every hurdle, into Naval Aviation.

Now, as the sun sat low near the horizon, we approached that historic place where provocations on December 7, 1941 resulted in all the turmoil, transformations, and concerted efforts we saw everywhere around us.

Our plane let down as we approached Oahu, West of Waikiki and Diamond Head. Our nose pointed toward the harbor where the deed was done. Slowly coming into view were the twisted steel superstructures of that string of battlewagons. Those reminders of our challenge sent a chill along my spine.

Soon we were down on the water at the far end of the inlet. We docked at the Pan American Terminal, took a bus to the Naval Base, and checked into a TOQ. There was still time to look for a ride to the mainland. At the Pan Am Terminal they told me it might be a long wait for air transport since too many folks with high priorities were using that route.

I took a ferry over to Ford Island and found the transportation office. They had posted a sailing for the next day: a baby flattop named USS *Hoggatt Bay* would depart mid-morning. It looked like the ride for me; would take five days, but I was used to this kind of vessel. I had gone over on a baby flattop, the USS *Barnes*, and would feel at home.

I had time to check out the famous city of Honolulu, so I headed back for the TOQ. Just as I got there I started to feel chills. I was baffled, thought those islands had a tropical warmth that lasted into the evenings and endured until morning. Out the window I saw everyone dressed in whites and khakis with short sleeves and loose collars.

Why am I shivering so hard? I'm so cold my teeth are chattering. This is not normal. I think I caught the bug. Malaria! That's it. That night on the rock when I got the mosquito net out and the little suckers attacked my hands and wrists. They got me and I got it: Malaria. So this is what it feels like.

I sought solace in bed, piled blankets on top of blankets to fight the cold and it worked. The chattering stopped, but soon I was burning up, sweating profusely, kicking off the covers. I tried to get comfortable, but it was hard. Finally, I fell asleep and slumbered the rest of the night away.

By morning the attack had abated, and I made it to my ship on time. I was at the rail on the starboard side as we passed alongside vacant docks and twisted steel, reminders of the enemy attack. On that day it seemed a sad and dreary reminder of death and destruction. The chills and fever were gone, replaced with feelings of depletion and emptiness. It must have been a side effect of those first symptoms.

The islands had disappeared behind, but I continued my stance by the rail near the bow. I remembered some sightings at sea from the earlier trip, in December, aboard the *Barnes* enroute to Espiritu Santo. I had an inordinate curiosity about the ocean. It entranced me.

Other guys on the trip sought out different forms of recreation. Most of them engaged in games of chance, poker games mostly, held in the ward room or the ready room. Others read or shot the breeze, but I stood alone by the rail gazing out over the ocean, noting the gradually deepening blue, watching flying fish leap up and splash into a wave of unbelievable translucent turquoise. You could still see their silhouette after they plunged below the surface. Every sighting seemed worth the wait, and so I went there every day and looked.

One day I was rewarded with a rare sighting. I happened to be focusing on the bow wave alongside the ship when a huge mass of protoplasm and tentacles came into view and slid right along the steel plates of the vessel. It was almost as big as a whale but didn't have any of its appurtenances: no flukes or fins or spout, just a huge mass with tentacles. I told some ship's crewmembers about it. The best guess they could come up with: a rare giant squid.

Then one night I had the duty. My job was to have an enlisted man make the rounds of the Corsairs tied down on the flight deck. (The *Barnes* was sailing, unescorted, at top speed with new fighter planes tied to its hangar deck and flight deck.) We were totally blacked out, making it hard to find other people on deck. I never did make contact with the crewman but I ended up on the bridge, the tall superstructure amidship, where I finally found another soul to chat with: the Officer of the Deck. We stood together, talking, leaning on the rail, looking into the dark waters below.

He said, "I recently put in a request for flight training and I... I... I... yi, yi, yi...!"

We both saw it at the same time: four phosphorescent streaks in the water aimed straight for the side of the ship. The OD swung around and dashed for the ladder. As fast as he could, he went down to and across the deck below to see if they came out the other side. I stood and waited. I had expected an explosion, but only silence prevailed and the whistle of the wind. It happened so fast that fear was only a passing thought, not a feeling.

Then I heard the sound of men ascending the ladder and saw their dark forms approaching. The OD introduced the ship's captain.

"What did you see?" he asked.

"Just four streaks moving abreast, straight for us, sir."

"Four abreast, you say. Couldn't have been torpedoes. Japanese subs can only fire two at a time. Probably small whales. They like to swim in formation like that. Quite a scare, though."

"Oh, yessir, a scare for sure," I agreed.

Interesting sightings along the rail rewarded my curiosity about the mysteries of the ocean—rewarding and worth the wait and the cold, but this return trip had a different feel. I had ashes to sift and it would take time; my attention had turned inward. I found myself reviewing my logbook and thinking about activities before the logbook began.

"Reporting in" that first day in mid-June turned out to be a glorious, glamorous entry into the wartime military experience. A Pullman car full of young healthy males had rolled into California from the Pacific Northwest. Our car had been shunted, along with others from around the state, onto the train ferry on the Bay's east shore. Our destination: San Francisco, that glamorous City by the Bay to the west. But why there? Our destination was Moraga, east of here. We were Naval Aviation Cadets, headed for a Catholic college, St. Mary's, east of the Bay, but here we were on a westbound ferry to San Francisco. We went under the new Bay Bridge, just completed three years earlier, and docked at the wharf where trainmen shunted our cars into a long building. Why were we in San Francisco?

We soon found out why: an officer in Navy blues moved down the aisle and told us to disembark. *Let's see. I think that means, "Get off the train."* So we did. A number of officers in dark blue uniforms with gold stripes on their sleeves and round white caps on top, had us line up along the curb beside a whole string of open convertibles, each with a pretty woman at the wheel. It began to dawn: we were about to take part in a publicity stunt. With our baggage loaded aboard a truck, we each settled our civvy-clad butts on the seats of the glamour cars.

A police escort led us to the bridge approach and as we crossed the bridge, eastbound with The City in the background, news photographers aimed their flashing cameras from nearby cars. We were escorted all the way to Moraga by police and newsmen. We were news because a Navy Pre-flight School was new, a nest where they got young fledglings in good physical shape, and got them used to Navy lingo and the theory of flight while their new training planes were being built in Dorothy's home state of Kansas.

We didn't wait. We got in shape. We really did. Most of the physical trainers from colleges all over the West ended up there in the program, many of them champions in their fields. We had Hank Lusetti, Frankie Albert, and others from Stanford, football players from Cal, and an Olympic champion gymnast, Lieutenant Keen.

As I lay there in my private quarters, high and close to the steel plates of the port bow, my gaze dropped to my *Aviators Flight Log Book* and I opened it again. I was thinking about the beginning entries, the initial flying lessons, first in a Piper Cub, an NE-1, an orientation hop with a Navy Lieutenant named J.F. Jolley, whose roundness of body and face went well with his name and disposition. It turned out that I couldn't have had a better instructor. His experience and self-assurance in the air along with his easy-going nature left a lot of room for student missteps. He had a higher tolerance for error than did most of his younger colleagues, as I soon learned.

My first few lessons went well because of my above average physical coordination, and I learned the necessary skills quickly. However, I found myself lacking in self-confidence when I had to take command of the plane, when I had to be right on top of every wriggle it made at slower speeds on the runway. I seemed to have a mental glitch at a crucial moment when the speed dropped and the craft suddenly became unruly and wanted to wander all over the runway, and Jolley would have to take over the controls and guide it to the taxiway.

I had learned the basic skills in the first nine flights in the Stearman biplane after two orientation sessions in the Cub. I'd at

October 1942

Date	Type of Machine	Number of Machine	Duration of Flight	Character of Flight	Pilot	PASSENGERS	REMARKS
12	NE-1	26335	1.5	"	Lt. Schutz		
13	"	26336	1.5	"	"	"	
15	N2S-3	05820	1.5	"	"	"	
16	"	05419	1.5	"	"	"	
19	"	05424	1.5	"	"	"	
20	N2S-4	30606	1.5	"	"	"	
21	N2S-3	05424	1.5	"	"	"	
22	"	05421	1.5	"	"	"	
23	"	05417	1.5	"	"	"	
24	"	"	1.5	"	"	"	
25	"	05419	1.6	"	"	"	
25	"	"	0.5	"	"	"	AX-1
25	"	"	0.5	"	Ens Smith	"	AX-1
30	N2S-4	27743	1.2	"	SELF	SOLO	

Total time to date,

learned to keep the wings level and hold a constant altitude above the rolling hills of south central Washington. I could do reasonably smooth banking turns onto outlying practice fields near the place where the Columbia River takes the big turn northward into Washington. I could even make good three-point landings—*carrier landings*, Jolley called them. Every landing we made, right from the beginning, would be a simulated carrier landing, the Navy way.

However, the goal for the Aviation Cadet was to reach the point where he could take the plane up alone. Solo, that is. At the end of flight nine, the make-or-break flight, Jolley just sat there with his head down. I had climbed out on the wing and stood alongside his cockpit waiting for his comments. *"Why is he just sitting there? This isn't good."*

When he looked up at me, I saw in his grey-blue eyes a troubled, concerned look as he said, "Cadet Morris, there is only one more thing you need to know and that is: you can fly this plane."

I was paying more attention to his demeanor than his words, but I thanked him for the encouragement. My heart started a downward spiral as I slipped off the wing and walked slowly toward the ready room. Visions of an ordinary seaman swabbing a deck flashed before my eyes. I could be doing that, like some of the guys who had "washed out" already, packed their gear and sadly walked away. *If I don't do something about my confidence tomorrow, I'll be on my way out, too.*

That night I hardly slept. Visions of that sailor swabbing the deck continued to haunt me. I turned over and over in my bunk trying to avoid them, wanting to focus on what I had to do the next day. I managed to get in a few troubled winks before the dreaded day dawned. When it did, I scanned the skies for rain clouds, hoping, praying for something, anything, to postpone this challenge.

But the skies were clear with only a few scattered clouds as I warily, tentatively made my way toward the flight line. I first scanned the flight board. *If your instructor is sick they'll cancel your flight.*

But no such luck. I found my name at once. This was it. The fateful moment had come, and now it was all up to me to decide my own destiny. I had to get in that biplane, taxi it to the runway, test the magnetos, then pull up to the take-off position and go.

Encouraged by my own thoughts, I hurried into the parachute room, chose one and hoisted it to my shoulders. As I did so, a mysterious flood of energy swept through my body and with it a feeling of self-confidence that told me all would be well. The normally heavy 'chute felt light as a down pillow as I hurried to my plane on the flight line. It was as if I had "put on" confidence when I hoisted the chute to my back.

Jolley sat waiting in the front cockpit as I easily climbed up the wing and practically vaulted into my seat with a feeling that nothing could stop me now. With the engine started, I signaled to a crewman to pull the chocks and I heard Jolley, through the tube, tell me to take it away. I taxied easily out of the parking area to the taxiway, on out to the runway where I checked the mags, then pulled into take-off position.

Until then there had been only that one comment from my instructor when he said, "Take it away, Cadet." He said it again as I pushed the throttle forward and responded with an "Aye, aye sir!" We were approaching the point where he usually had to take over the controls when the torque of the prop pulled the plane off to the left, but this time I had complete and utter control of the Stearman as we became airborne. I could hear chortles and chuckles from Lieutenant Jolley, living up to his name, building up to a real guffaw as he vented the great relief he felt at having his student show success.

"What did you do last night?" he shouted. *I think he's implying something uncivilized.* I looked up in the mirror, at his broad grin. *If he only knew!* I grinned back. *He'll never know.*

The rest of the flight was a breeze. At the small field I shot three landings. The first was a solid three-point, full-stall landing on the near end of the runway, no wandering. I was in complete control of every subtle little move and took off immediately. Jolley had not once touched a control.

On the second landing, I came in a bit high, stalled high, and we bounced back into the air. But I was on top of the situation and caught it with a short burst of throttle that let it down to a gentle landing. We ended up too far down the strip to take off again. I made the decision to stop, turn around, and taxi back down to the takeoff end. I had absolute control of the plane at all times, which pleased my instructor. He said things like, "You'll like it in Corpus. It's my home town." He meant Corpus Christi, Texas, the location of the next stage in flight training at the Naval Air Station there.

I pulled off another slick landing, a three-point Navy job at the end of the practice strip. The test was over and I got a thumb's up! En route to the main base at the Naval Air Station, the Primary Base at Pasco, Washington, Jolley told me he felt so confident in my flying that he could get out after we landed and let me solo right then, but Navy Regs would not allow it. I needed a check flight from another instructor.

After we landed, he told me to wait by the plane and he would send someone out to give me a check ride. He found Ensign Smith listening to a ball game in the instructors' ready room, who was willing to give a half-hour for a quick check flight. I made three good landings, which satisfied him enough to give me a thumbs up. *I get to solo, to take off and fly alone!* The vision of the seaman swabbing the deck had faded. The fear of washing out had abated for the time being.

What a thrill! An unforgettable experience to soar up there all alone, to get past that hurdle that seemed so formidable the night before. What a delight to be up among the clouds in that little yellow biplane as it responded to my every touch on its controls, to be in complete command of all maneuvers and to bring it down to a three-point, full stall landing at the end of the runway all by myself. To be a *flyer*.

In my logbook I saw that my solo date was October 25, 1942. I made a note of it, then thumbed through the pages. I thought about other challenges and difficulties I had to meet before I got my gold wings: night flights, aerobatics, formation flying, instruments, seaplanes, navigation, and catapult takeoffs. Those last flights brought back a memory of another incident when I was threatened again with the possibility of a washout.

It happened eleven days before graduation at Corpus, on the day my friend, Bill, graduated and was married in the little base chapel. He wanted me to be there but I had a flight and by the time I had finished I didn't have time to log out of squadron headquarters. I ran straight from my plane to the bus but when I arrived at the chapel, I found it empty. Disappointed, I started back for the squadron ready room but somehow I ended up in one of the barracks buildings saying goodbye to some others who had just graduated and never did get back to my squadron. I just went to dinner from there. *Looks like that long 2.7 hour flight I took on June 12... in my logbook.*

Meantime, the skipper at the squadron had called a muster so he could give a lecture. When Cadet Morris failed to answer, he informed the other cadets that I was in trouble, and that he wanted to see me first thing in the morning. Next day when I went in to see him, he informed me I had committed a Class A offense: absence from post of duty and it could mean a washout. There was that horrible word again. It brought a devastating feeling—and for something that didn't seem that serious. *I simply missed a lecture. Couldn't I make it up somehow?*

He didn't ground me. I found my name on the flight board every day, but this affected my flying, this thing hanging over my head. The same day the skipper told me what he was charging me with, I had a flight that I came close to messing up, badly. It was

June 1943

Date	Type of Machine	Number of Machine	Duration of Flight	Character of Flight	Pilot	PASSENGERS	Dual Time	Solo Time	Pass REMARKS
1	05243	5390	2.3	BW	Cpl. Cochran	Self			2.3
1	"	5390	2.2	"	Self	Cpl. Cochran		2.2	
1	"	5601	1.3	"	Cpl. Cochran	Self			3.6
2	"	5405	1.7	"	Self	Cpl. Cochran		3.9	
2	"	5405	1.7	"	Cpl. Cochran	Self			5.3
3	"	5422	4.1	"	Self	Self		8.0	
5	"	09494	1.3	B	Lt. Krizak	Inst.	1.3		
5	"	5664	1.0	"	Self	Solo		9.0	
7	"	5608	4.3	"	Cpl. Christensen	Self			9.6
8	"	5506	3.9	"	Lt. Ray	Inst.	5.2		
9	"	5528	3.9	"	Self	Solo		12.9	
10	"	5631	3.7	"	"	"		16.7	
11	"	5506	2.7	"	"	"		19.1	
12	"	05412	2.7	"	"	"		21.8	
13	"	09489	1.1	"	"	"		22.9	
15	"	05692	2.1	"	"	"		25.0	
15	"	09465	1.0	"	"	Cpl. Christensen		26.0	
15	"	09485	1.2	"	"	Solo		27.2	
17	NSN3	4482	1.5 BY	Lt. Bayer	Inst.	6.7		
17	"	4482	1.6 BY	Self	Solo		28.8	

Total time to date,

something like a solo strafing flight that I flew from the main strip at Corpus in a land based OS2U Kingfisher. The field was configured like spokes in a wheel and takeoffs were made from the hub. You just taxied down a runway not in use.

I felt so distracted by the events disturbing me that I proceeded to make a series of errors before I finally got in the air. I had taxied cautiously out to the takeoff spot, zigzagging all the way to make sure the path was clear ahead. I called the tower for clearance to takeoff, checked the two banks of magnetos, then wheeled into position and started down the runway. At about halfway down I noticed the engine seemed to lack power; that there was barely enough speed to get off the ground. I chopped the throttle, aborted the takeoff, and rolled to a rough stop near the end of the runway. To return to the flight line I would have to taxi back up the runway—I felt compelled to call the control tower and request permission.

I picked up the mike, pressed the button, and started to talk but it sounded flat, not picking up in my earphones. Frustration built into anger and I said, "Sh--, what the f--- is wrong with this goddamn thing?" My eyes were focused on the instrument panel as I continued to curse into the dead mike when I noticed, out of the corner of my eye, an orange light flashing with each word I pronounced, each four-letter invective that spewed from my mouth. *My god, my cursing went out over the airways!* Everyone in the traffic pattern; everyone tuned to this frequency, the women operators in the control tower, all heard my curses. Then I noticed that the line from the radio to my earphones had come loose in my mad scramble to bring the plane to a halt. What I had thought a dead mike was dead earphones. Now I was really in trouble.

I plugged the phones back in, called again for permission to taxi back up the runway and zigzagged all the way back to the flight line. A mechanic hopped up on the wing and asked, "What's the problem?"

"I dunno. It just lost power on the takeoff."

He poked his head in the cockpit and scanned the instruments and controls. "OOOH... Looks like the magneto switch is in the wrong position. You tried to take off on only one bank."

Chagrined, I slapped my forehead. "Thanks." I flipped the switch to BOTH. He pulled the chocks and off I went again, zigging 'n zagging along the taxiway to takeoff position. This time I completed the flight but not without concern about my performance on the radio. My worries were compounding themselves. *The Skipper wants to wash me out and I just swore over the public airwaves. What next?* I still had some night flying on my schedule along with four catapult shots and precision water landings. *Lotsa luck!*

Back at my barracks I got word that my Battalion Officers wanted to see me right away. I thought they must have word about the Skipper's charges against me or maybe something about swearing on the airwaves. I hurried anxiously into the headquarters office. There I was met with friendly and sympathetic words of encouragement. These officers were not flyers, not Navy career men but were intelligent, professional men, reserve officers. I had a clean record, they reminded me, and I was a cadet officer, a two striper. They were clearly on my side.

They said the Commander had tried to get them to charge me with a Class A offense and they informed him that what I had done didn't fit the category. They told me they could charge me with a minor infraction that would have me, as a cadet officer, serve my punishment by sitting a few hours in my room.

And nothing ever came of the cursing incident. Maybe they really didn't know who did it.

This good news changed my entire attitude from night to day, from nightmare to mindful alertness. And no news about the radio goof-up was a relief, too.

I saw that the next six entries in the logbook were all on 21 June, all in floatplanes, four at the catapult.

Oh yeah, those four catapult shots: first in the back seat with

June 1943

Date	Type of Machine	Number of Machine	Duration of Flight	Character of Flight	Pilot	PASSENGERS	Dual Time	Solo Time	Pass Time	REMARKS
21	0S2U-3	5628	0.2	C	Col. Christenson	Self			6.9	Catapult
21	"	5628	0.2	"	Col. Christenson	Col. Christenson			218	
21	"	5628	0.2	"	Self	Self		29.0	9.6	
21	"	5628	0.2	"	"	Self		29.2	9.8	
21	"	5596	2.1	BY	Col. Donaldson	Self			10.0	
21	"	5596	2.1	BY	Self	"		31.3		
22	"	5389	2.3	BY	"	"		33.4		
								35.7		

Total time to date,

another cadet piloting, then in the pilot's seat with the same cadet as passenger; then in the pilot's seat with another cadet—a mere fledgling—in the rear, and finally the last, in which the fledgling got his shot in the pilot's seat.

Each shot put us in the air in three seconds, from a dead standstill to over 60 knots, almost instant flight. It was important to brace yourself against the seat back and headrest if you didn't want your head to bounce around like a ping pong ball. Once in the air, we circled into a traffic pattern, then made a dead-stick-precision-landing next to the dock and under the hoist. These were simulated cruiser or battleship landings, the way they make landings at sea. The guy in the back seat had to get out on the wing, grab the hook, and put it through the loop in front of his cockpit with the plane still moving (there are no brakes on seaplanes). Then they hoisted it back onto the catapult.

That last catapult entry had my attention. When I climbed down from the rear cockpit that day someone came to me with a message: The Commander wants Cadet Morris to report to his office on the double.

I climbed aboard a nearby bus with fear in my heart. This guy was still scaring the hell out of me in spite of the reassurances of my battalion officers.

At squadron headquarters I walked slowly toward the Skipper's office.

I knocked, with some trepidation.

"Come in."

I entered, with serious puzzlement on my face, and said, "Cadet Morris reporting, sir."

"Oh, yes. At ease, Cadet. Have a seat."

He had a question for me. "How many flights do you have left to finish your training?"

"Just two nights of night flying, sir," was my reply.

"Welllll... they need pilots out there. It would be a waste to throw away expensive training like that. I'm going to let you go. But be careful from now on, this will be on your record. This mistake you made."

"Oh, yessir, thank you sir."

"Oh, and by the way, I'd appreciate it if you'd not spread it around that I was easy on you."

Through a big broad smile: "Aye, aye, sir. Thank you very much, sir."

He dismissed me and I walked down the hall toward the exit, feeling liberated from that oppressive, smothering feeling that had possessed me for the past week. I didn't feel grateful to the Skipper for this release. I wouldn't have to keep silent about his being easy on me. My graduation five days later would boldly tell the whole tale.

CHAPTER 7

ROUGH SEAS AHEAD

Except for the waves of malarial fever and chills that swept through my body three days later and the mountainous waves of the Pacific that broadsided our flattop on the last night before landfall, the trip to San Diego was relatively uneventful. The flood of memories that came at the height of my fever and the furious storm outside, however, were notable.

The steamship bucked furiously and made my stomach hurt. I woke in the night with an urge to throw up, a violent new experience for such a young and inexperienced traveler at age two.

I was, only then, awakening to my existence in life with the retching of my insides, outpouring the vile contents of my stomach. What a way to become suddenly aware of life.

I called out to my mother but she already knew I was seasick and ministered to my needs with her comforting presence, cleaning me and my bed, reassuring me. "It's all right. You just threw up. You got seasick from the rocking boat."

"Did Lois throw up?" I looked over toward her bunk.

"No, but she might." Momma replied.

During the night we had passed Glacier Bay and sailed out into the wild and stormy seas of the Gulf of Alaska. My mother and sister and I were on our way to join my dad in a remote area of Alaska's wilderness: the Wrangell–Saint Elias Mountains.

I remember nothing that led up to that traumatic moment of my childhood. Suddenly I was there, throwing up in my bunk and Momma, comforting me.

The next morning we disembarked at the docks of Cordova into a strange new world with noisy autos and horse drawn carts. We had to go from the ship to a taxi, which was a lot smaller than the ship but bounced along in much the same way.

"Where are we going?" I asked.

"We have to go to the train station," Momma said.

"Is Daddy there?"

"No, he's in Kennecott. We have to take the train there."

At the station we found that the train wouldn't leave that day but we could take a thing called a jitney, a big touring car with isinglass curtains and railroad wheels that ran on the tracks.

We went all the way, bundled up against the cold, through the day and into the night on that jitney with its curtains flapping in the breeze, our luggage fastened on the rear and in the running board rack. It was early summer and didn't get very dark in this northern latitude. Inside the curtain, I slept and didn't see the dramatic landscape we went through, but on our arrival I climbed out into the long shadows of huge mountains, their snowcapped peaks reflecting the pink glow of the low-lying sun.

There were impressive buildings all around, most of them painted deep red with white trim; some were part of the mill, others apartment buildings. This was Kennecott Copper Mine and Mill with its own town built right in. My dad worked there. He was almost a stranger to me. He had missed my first and second birthdays. We would have to renew our relationship, which we did, in extra short order. I remembered his thick black curly hair and his strong arms.

We would be living in tight quarters. Until an apartment was ready, our home would be a tent house. We all helped unpack the luggage from the jitney to our new abode where we settled in and did a circle hug. We were a family again.

That return to the States was not the great joy I had expected. The malaria and my wounds and perhaps my recent survival experience all weighed heavily, unexpectedly, on me. I felt sad, depressed and withdrawn.

I thought about Rosalind with trepidation. *How could I expect her to honor our betrothal when I was no longer the same man who courted her and gave her a ring? I'm a fallen warrior who could lose more of my eyesight. The malaria could diminish the good health I have enjoyed in my young adulthood. Life just won't be the same as before. I really don't know what to expect. I do feel lucky to be alive, but can't seem to shake this depressed, worthless feeling; it's like being unemployed.*

When the California coast came into view I watched with detached interest, as the golden cliffs grew larger. The ship made a turn to the north and we sailed along for another hour with no harbor in sight. During the night, big waves blasting against our port bow along with the heavy wind had blown us more than twenty miles off course and we were still off the coast of Baja California.

It was almost another hour before we reached the opening to San Diego Harbor, between Point Lomas and the Naval Air Station at North Island. As we glided in close to the docks, warplanes landed at this very field where I had flown in just three months ago on my last flight, before we shipped out from here on another baby flattop.

I had made that flight to North Island and back to Mojave the day before I sold my Willys, the little black coupe that had served me so well on those hundred-mile-one-day-liberty-trips from Mojave to L.A., for dates with Roz.

Dec-43

Date	Type of Machine	Number of Machine	Duration of Flight	Character of Flight	Pilot	PASSENGERS	REMARKS
1	F4U	02893	1.5	F	MORRIS J.G.		Dummy
2	F4U	17800	1.5	A	"		Both Types
2	F4U	02442	1.5	A	"		"
3	F6F	13074	1.5	F	"		Dummy
4	F4U	17800	2.5	L	"		Wayne - Bruce Cobb
4	F4U	17800	2.8	L	"		Bruce Cobb. Wayne
6	F6F	13062	1.0	A	"		Both Types
6	F4U	17471	1.5	F	"		Firing
7	F6F	13188	1.5	F	"		"
7	F4U	17794	1.0	A	"		Both Types
8	F6F	13188	2.5	A	"		"
8	F4U	17630	2.5	A	"		"
13	F6F	13188	1.5	A	"		"
14	F4U	02649	1.5	A	"		"
14	F6F	13188	1.0	A	"		"
15	SNJ		1.5	L	"		McADLE - SAN-DIEGO
15	SNJ		1.5	L	"		SAN-DIEGO - MOJAVE

Total time to date.

On that day, December 16, 1943, Roz and I had to take care of business. The squadron would be shipping out before Christmas. That meant store your car or sell it. I had decided to sell. It needed a little work, had blown a hole in its manifold gasket and it whistled and popped, but I knew it would bring enough money for a nice gift for Roz.

I let her out at Bullocks Wilshire, that elegant art deco store with the tower and the black and turquoise tile, facing Wilshire Boulevard. I found a used car lot on nearby Olympic Boulevard where I unloaded "Little Willy" for a cool 250.

We looked at wristwatches. I wanted to find something as classy as Roz and I thought I had it when I found a neat little art deco number that looked good on her wrist and fit in with the atmosphere of the place. She liked it too, but she wanted to do some comparison shopping just to make sure.

She said, "Let's go over to Howe's Jewelry... across the street. There might be something different there."

I agreed, "We can always come back for the watch."

I'll always remember that watch even though I never saw it again. At Howe's, we looked at watches heavily embellished with gemstones. I felt that she was too classy for those fancy pieces. I told her she was more the art deco type. By accident we found ourselves staring into the engagement ring display. I said, "Wanta try some on?"

She said, "Should I?"

I said, "What the hell!"

That was how we got engaged at Howe's. The jeweler, Mr. Howe himself, happened to have a box of Sees Chocolates and he gave us our first engagement gift.

Then we had to go find "the girls" from "The Club" to break the news, pass out candy, and show off the ring, a respectable medium sized solitaire with two baguettes. "The Club" was a group of young women who had been getting together monthly since high school. Most of them had known each other in elementary school. Having thus ritualized the event adequately, we hied off to the Coconut Grove to celebrate to the music of Freddy Martin. He had become our favorite bandsman when we fell in love dancing to his tunes from Oklahoma: "People Will Say We're In Love" and his theme song, Tchaikovsky's Concerto, "Tonight We Love".

As I gazed into the sky, dreamily watching the war birds droning overhead, I heard the faint sounds of music. It was coming from the dock below, from a military band playing. I saw flags and a color guard. It seemed ceremonious.

There must be a VIP aboard or something. Maybe they are going to have a parade. Maybe they are just practicing.

I couldn't make the connection that they could be playing for us sailors, marines coming home from combat. Something inside me rejected this.

I came home too soon. I goofed. Save it for the real heroes.

Instead of being elated by festivities on the dock I felt dejected, sad. I really wanted to be back in the South Pacific with my squadron, my fellow Marines.

CHAPTER 8

THE BIG HOSPITAL

I climbed aboard another of those omnipresent olive drab ambulances with a big red cross on a white circle, bound for the Naval Hospital in Balboa Park. We crossed the bay aboard the Coronado Ferry, drove right through the center of San Diego's busy downtown, alive with men in uniform, mostly sailors, walking in every direction on sidewalks, in crosswalks.

The monumental hospital loomed on a rise behind a vast field on the edge of the urban area. Its beige walls with red tile toppings and tall bell tower were reminiscent of California missions. Marine guards stood aside to let us through the gate on our way to the admission office.

After I got my room assignment, I wandered around to acquaint myself with my new surroundings. Later in the day I met my new doctors. They had sophisticated, state of the art instruments: ophthalmoscopes, slit lamps, and x-ray machines. They spent a long time looking into my eyes, saying a lot of "hmmms," making notations on charts, and putting drops in both eyes to dilate pupils. Before I left, they handed me a bottle with a dropper and told me to put drops in both eyes every morning, come see them every day, and avoid excessive physical activity. On the way back to my room I looked at the bottle. The label said Atropine. It would be my constant companion.

As the end of the day approached I felt more and more restless. I hadn't called Roz yet. The reunion wouldn't be easy. It wasn't the same. *How will she respond to a wounded warrior?* I sensed that I shouldn't expect her to honor our engagement since I wasn't the same as when I left: a flying officer with a promising future. Of course, combat flying didn't exactly guarantee a future at all...

I'll bite the bullet, call her, and we shall see. So I did. She was home. She answered, "Hello?"

"Hello, it's me."

"Jackson! Where are you?"

"San Diego, the Navy Hospital."

"Oh, golly, I want to see you right away, darling."

"You're sweet, but it's the middle of the week. What about work? And you would have to take the train."

"Oh, I quit my job, and I could get Louise to drive me. We'll come tomorrow afternoon."

"You qui... ?"

"I'll tell you about it when we get there. I want to know all about what happened to you. We should arrive around five."

"Okay, see you then."

"You sound kinda down... what's wrong?"

"Oh, I dunno. I have malaria... seems to make me kinda gloomy or something. Anyhow, see you tomorrow... love ya over."

"Roger, over."

We had adopted the military communications lingo as our own personal code for saying goodbye. As long as it ended with "over", it meant our relationship was ongoing.

Something about our closeness warmed me. Good for the spirit. Sweet memories bolstered morale. I remembered how her image in the clouds above the jungle river inspired me, yet something else prompted my concern, an issue aside from those of my physical condition.

All through dinner, I thought about her, the good times we'd had, her smile, the laugh with the sounds of a gentle brook. Those happy, carefree dates when we danced, dined, went to shows, strolled in city parks, drove through Hollywood, Beverly Hills, Santa Monica, Wilshire Boulevard. Visions of those vibrant, spirited days tended to cover—even smother—issues too close to thoughts of marriage.

I don't want to think about it, I tell myself.

But I am thinking about it. It's on my mind.

Okay, I kinda left it up to her and she never said a word.

But that doesn't make it go away.

But since she said nothing, it must be a non-issue with her. Besides, her brothers and her sister have all married Gentiles or non-practicing Jews. It looks like they are cutting all ties at this generation.

Let's see, I wouldn't have suspected she was Jewish if I hadn't received that anonymous note telling me that if I got too serious about the girl I'm seeing I could end up "celebrating Yom Kippur in September."

Well, it shook me... at first, but I'm in love with this girl, dammit and I don't care!

So, I can just bury the issue? Cover it up and never talk about it?

Wellll...

And what about Mom and Dad? Especially Mom; she already showed how she felt when they first met last summer.

And there was that woman visiting Roz's Mom that day last fall who looked at me and said, "You don't look Jewish. You look Irish. Did you say your name was Morris? That's a good Jewish first name but... I dunno..." and Roz tried to shush her but she just rattled on, bringing up issues nobody wanted to hear.

Isn't this the age-old dilemma? The Montagues and the Capulets, the unblended melting pot: differences stirring up the mixture, elbowing and shouldering and fingering out the ones that don't fit in because of skin color or race or diverse cultural habits. Prejudices, biases, and downright segregation accentuated by the use of derogatory names: Colored People, Jews, Mexicans, Chinese, Italians, Irish–who were often called Niggers, Wops, Kikes, Spicks, Chinks, Micks (which includes a good part of my ancestry)–and of course, our current favorite disparaging names, Japs and Krauts.

So you look into the melting pot and see all those souls struggling to be on top and needing so badly to find scapegoats and ways to feel superior. You wonder what Jesus would think of all this. Would he take sides, find ways to fit in with the top dogs, use derogatory expletives, gang up on underdogs? Whenever I come up on a dilemma where there is a question of personal honor, I use the Jesus paradigm.

And of course Mom has been a big influence in my life. She's the one who took me to Sunday school, in essence, introducing me to Jesus who, by the way, also was a Jew.

I'm in an engagement with a Jewish girl whom I know Mom disapproves of... a city girl, she says. Where does my devotion and loyalty go? If I invoke the Jesus example my fiancée is innocent, not to be spurned... but... what about the rap about honoring your parents... she's been my Mom a long time... but that narrow escape from death I had, changed me. I'm not her little boy anymore. I've experienced a rebirth of sorts, and gained a degree of omnipotence or invulnerability and autonomy.

I was having this heart-to-head talk with myself, feeling tired, hanging out in the lounge, thinking about going to bed. My roommate snored loudly and the only way I could get to sleep was beat him, I mean, start making zees before he did. It was not easy to hurry-up-and-sleep. And that night was especially no exception.

Late the following afternoon, they arrived: Roz and Louise. We met in the lounge, the first familiar faces I'd seen in more than a month. What a sight to see. We embraced, happily together again. Still, an underlying current of sadness pervaded. I saw it in her pretty eyes. I felt it in my bones. I still had this question on my mind about expectations, about changes, commitment, still wondering—*is the cultural difference really that different?*

I told my story, the abbreviated version, because they couldn't stay long; they had a long drive back that night. Roz told her story—not a real explanation of why she quit her job—that she had been ill for a while.

As dusk fell we walked out the main gate, toward the parking lot. Louise bid me farewell and left us alone to talk. We walked across the street along the fringe of Balboa Park. Roz sensed my hesitancy, tentativeness in my bearing, a subdued manner. With something to say, but not saying it.

"What's wrong?" she asked.

I didn't want to make any rash statements so I blamed it on the malaria. "They can't get a positive smear so they haven't given me anything for it. Those fevers are really hot. The chills are no thrills either."

"Please call and let me know how you're doing. I'm worried about you. I can visit on weekends."

That old feeling was seeping back into my being. "Yeah, that would be great. We can go out on a date."

We embraced. A farewell kiss and yes, the familiarity of our closeness brought promise of renewal. It was spring again. *Thank God... our first Spring together. We may make it as a couple after all.* That dark cloud of doubt faded to a light haze. My spirits were lifted as I strolled back across the grounds with a lighter step. I'm glad I didn't say anything.

Attention to my eye began in earnest by medics in concert; besides the Navy doctors there were civilian ophthalmologists, older men eminent in the field, who came in weekly to lend a hand as volunteers. They spent long minutes peering through their instruments, seemingly captivated by unusual conditions in the posterior chamber of my left eye.

They told me about a scar they saw on the edge of the retina. They warned me not to engage in activities that could jar the retina and cause it to pull loose. Scars like that sometimes resulted in detached retinas. They could be serious and difficult to treat.

They ordered x-rays of the left eye in hopes of locating the fragment that did all the damage. Aside from the scar, they found that it had caused a "traumatic capsular opacity" on the back of the lens. They described it as a form of cataract, an irreversible condition.

When I went in for the x-ray, they told me they would have to install a contact lens. I'd never heard of such a thing. Since it was a magnifying lens and quite thick it would require a local anesthetic. A number of shots were taken from different angles.

The x-rays revealed three pieces within the orb. This was no surprise to me since small particles had been working their way out the side of my nose. The largest, in the eye, lay in the ciliary body. It had plunged into the eye, slashed the edge of the retina and caused opacities on the lens. The ciliary muscles are fragile filaments that support the lens and cannot be cut; the fragment could only be removed if it was of ferrous metal and responded to the forces of an electro-magnet.

The Doc explained that they could cut a flap on the surface of my eye, the lower part below the iris, very close to the location of the particle. That would give a powerful magnet a good chance to pull it out.

The surgery was scheduled for early May. In the meantime my malaria was heating up; every three days it could be expected to flare up right on schedule. I wondered why it wasn't being treated. I asked one of the doctors if there was something they could give me. He told me they hadn't given me medication for the malaria because they had failed to get a positive smear as yet.

I'd called my parents and they were planning a trip to San Diego. They just had to see me as soon as they could arrange it.

I'd told them as much as allowed about my ordeal in the combat zone but they had to see me and touch me. They went through a lot of stress and sorrow when that telegram came. Their pain wouldn't be completely gone until we were together.

Roz and I were in touch by phone. She had a new job working as a dental assistant. Our relationship was warming up again. I told her I might be able to make a trip to L.A. when I could get some relief from the malaria.

Then, one night, it got worse. My temperature, after the chattering chills, rose to new heights, accompanied by nausea that resulted in vomiting. A corpsman took a blood sample and contacted the doc, who ordered medication. They got a positive smear and the Atabrine administered at the height of the fever wiped out the malaria forever. My temperature had risen to over 104°.

A week later I felt so good I took a trip to L.A. to see my fiancée and my good friends, the Whites: Nellis, Bob, and their little son, Bobby. I had known them since my first week out of flight school when my friend Jack and I responded to their listing of rooms for service men that we found at the Officers Club, where I met Roz. It was more than just rooms: we were always invited for meals and companionship. It became a home-away-from-home for a number of men, mostly us flyers and some infantry officers from the Marine base at Camp Pendleton. They were really glad to see me. There were still noticeable facial blemishes, scars, and a paleness about me. It was a kind of payoff for their contribution to the war effort to hear the survival story from one of "their boys".

The wonderful part of that visit for me drifted in from the kitchen—the absolutely divine smell of a home cooked meal, followed by its very merry consumption at the table with Roz amid a circle of good friends.

When the time had come to depart, I stood and felt a slight chill. Nellis noticed a wanness about me and commented, "You look so pale. Will you be alright?"

"I feel good. I feel a wee bit cold, but it's nothing compared to the chills I've had for the past month." I was, at that moment, feeling the last vestiges of the disease.

"Soon we can go on a real date together," I said to Roz as I let her off at home on my way in the taxi to Union Depot. "We'll go to the 'Grove'... like old times."

"Sure, darling, take care of yourself now."

The next day I contacted my general practitioner again because I had noticed my left wrist still hurt when I tried to get around or use it for anything. He checked it with a left hand grip and found I didn't have much strength there, and decided it should be x-rayed. It had been three months since the crash. A sprained wrist shouldn't take that long to heal.

The x-ray clearly showed a positive hairline fracture in one of the small wrist bones. I would not get a grip back unless they immobilized it for six weeks. I went to the orthopedic department at once to have a cast put on my lower left arm. This would slow me down a little, along with the threat of a detached retina. I tried to learn to baby myself a little with all those debilitating little complaints, punctuated by an occasional shooting pain in the left eye.

About a week after that Mom and Dad arrived for a short visit and at the same time my sister, Lois, showed up; a pleasant surprise. They were—pardon the trite phrase—truly a sight for sore eyes. My parents had visited me at Mojave the previous summer but I hadn't seen Lois for a couple of years. She was doing her part in the war effort as a civil draftsman at the Naval Air Station in Alameda. We had a joyous reunion... with a group hug like the one we had in Kennecott twenty years ago; only this time we were all adults, hopefully.

I told Dad about how I used the knife he made for me, carried it in its sheath through the ordeal, used it to cut through brush, the taro root, plantains, and opened a coconut. He was proud to be a part of it. I could see it in his face.

The visit was brief. They all had to get back to their war jobs and I felt better even though I had surgery to bear in a week. I told them I would come home on leave as soon as I recovered from the surgery.

They would not be giving me a general anesthetic for the operation. I had to stay awake and alert to assist in my own surgery: I would have to raise my eyes to expose that part normally hidden by the lower lid. In other words, keep looking up and hold still while they cut and sewed. *Ouch, this sounds difficult.*

When I tried to get to sleep that night, I thought about the first and only surgery I'd had, at the age of sixteen, a tonsillectomy. "Breathe deeply and count backward from one hundred," a nurse said.

I think I made it to ninety-six as I breathed in the sickeningly sweet ether. I woke up some time later to sounds of whistling—no particular tune—in fact, somewhat off key, like a kid whistling in the dark. My eyes were closed but I was awakening to the realization that the whistler was me. So, I stopped. Then I could hear a man's voice... in a story telling mode. He, presumably my new roommate, was relating to another roommate a story of some kind of concoction he had to take in the distant past, one of those home remedies, for a severe case of puppy love he had contracted as a teenager. It was something like sulfur and molasses. It seemed to have worked.

The realization that an acute case of coincidence had just taken place made the slits of my eyes slowly widen and slide sidewise to sneak a glance at the storyteller, a man older than my father.

Will I be telling the story of how my love-sickness was cured when I'm as old as he? His parents treated his with a dose of old-fashion folk medicine... mine was extracted with a set of tonsils.

I never told my parents about the girl I met on the summer break at the seashore. It was just my mother and me at Uncle

Kelly's beach cabin. Dad worked the week and spent two weekends with us. Lois, freshly graduated from high school, had a job. I had my bicycle and solitude on the sand; too much solitude, I realized as I pedaled aimlessly along. It would be really nice to make acquaintance with that cute blond I saw walking the beach every day with what looked like a younger sister.

How can I meet her? I have this shyness about me ... I don't just go up to her and say, "Hi. Let's get acquainted... or, how's tricks?" I think she's noticed I've been eyeing her, though. She sort of smiles back. Two days have already passed. The week's almost half gone. If I don't make a move soon, the week, and the chance to meet a doll will evaporate... but how?

Well, let's see... I could just fall down in front of them. I know how to fall without hurting myself... make it look like my wheel caught on something... tripped me up... yeah, that'll do it. It's now or never.

So I did it. Just like planned and it probably looked really stupid but I found myself at the feet of this cutie, looking down with concerned eyes above an amused little smile. Her sister giggled, but we were soon walking together, chatting about schools, hometowns. She hailed from Newburg, a few short miles from the place my ancestors had settled after their trek along the Oregon Trail in 1853.

All day we walked and talked, the three of us, and the next two days we met—the one bike spill was enough—and strolled and chatted. I had fallen for this girl literally and figuratively. Then Saturday came and the family: Dad, Aunt, Uncle, Cousin, friends, and with them a change in my routine, but she was constantly on my mind. Back home in Vancouver, I could think of nothing else.

Two weeks before school started a heat wave found me in the basement trying to keep cool, trying to deal with feelings of sadness. Mom caught me moping around down there, asked me what was the matter. I laid my head on her shoulder and started sobbing... but I didn't tell her what was bothering me. Never did.

She said, "Must be your tonsils. I'll call the doctor right away."

So, that's how my love sickness was misdiagnosed, but strangely enough, the treatment worked. Not only did I recover, but on my return to school I felt so exhilarated I went out for football. I really did have bad tonsils and a tonsillectomy beats the hell out of sulfur and molasses.

Seven years later I assisted in my own surgery as the houselights dimmed and a bright spotlight came up. It focused directly onto the exposed, wounded orb as the torture began. First they applied local topical anesthesia before putting on the clamp—the device that holds the eyelids open and steady... *ouch...* to keep me from blinking... *oooh.*

I could only see the bright light and moving shadows of attending medics. A small dark form approached my eye and I felt more drops being added. Since I could no longer blink, someone up there now served as my tear master to keep the eye moist.

A bit more local anesthetic and then they came in with the needle, a thin shadow at the end of a hand shadow, to deaden it even deeper in preparation for the first cut... *ouch. Oooh... he is really pushing it in... and I can feel pressure from the inside too.* The bout with pain was soon over as the analgesic took affect and they were ready for the first cut...*oooh. Another shadow. It's the knife... it's cutting... no pain, but I feel the pressure and the sound... like cutting into a carrot.* They made two incisions, forming a flap, and held it back with small clamps.

The bright light dimmed as a dark shadow approached... *a big shadow... must be the magnet with its moment of truth.* Nearly all the light blanked out as they got it in position, then called for power. A switch snapped. We all held our breaths. I listened for the sound of metal striking the magnet, for sensations inside my eye. Nothing. The only sounds were sighs of disappointment around the table.

All they could do now was, *ugh... sew it up.* The mere thought of someone doing stitchery on my eyeball made me cringe. Now I had to actually endure this procedure. *And I can't even shut my eye. Of course there is no pain but I have to tolerate the pushing and pulling. Ooh, the pressure.*

Then they finished. The clamps came off and I could blink again. *What a relief... and that bright light goes out at last... whew.*

A nurse prepared me for a shot. I asked the doctor what it was for. He said, "It's dead typhoid. It will produce an artificial fever that will hold off the threat of infection."

"Is it anything like malaria?"

"Oh, no. It will be mild by comparison. I understand your concern. You had quite a bout with malaria, but believe it or not, it may have saved your eye. The fever I mean. We found evidence of infection in your wounded eye. We carefully watched your good eye... sometimes infection can travel from one eye to the other... if we had found a trace of infection we would have had to remove the bad eye to save your sight. So your malaria was actually your friend... may have saved your vision."

Unknown to me, Roz was making a surprise visit. She knew about the surgery, but not the date. I had failed to tell her about the cast on my arm. As she came through the entry into the inner quad she saw a patient with a huge bandage on his head and a cast on his arm. He was in a wheel chair in the care of a red-haired nurse.

She thought, *another poor, dear casualty of war. He could be someone's husband, son, or brother. Or he could be someone's boyfriend or... say he looks a lot like...like my boyfr... It is!* "Jack! What happened to you?"

"Oh, Roz. I didn't know you were coming today. What a surprise..."

"Forget my surprise. What about yours—what happened?"

"Well, there was this big bang and everyth..."

"No, no. I mean... look at you. In a wheel chair, all bandaged up."

"Oh, that's right. You didn't know I was having surgery today. The piece didn't come out. They put this cast on my arm because they found a broken bone in my wrist."

"Oh, I'm so sorry."

"Never mind, darling. It will heal in six weeks and the little piece of shrapnel has done all its damage. I'll be ready for Freddy in a couple of weeks. We'll go to "The Grove".

She stayed the afternoon but I was sort of wrung out and not very good company. I fell asleep, beginning my recovery from the surgery, and she slipped out quietly.

The next day at the eye clinic the red-haired nurse asked me, "Who was the pretty young woman that came to see you yesterday?"

"That was my fiancée," I answered proudly.

Her only response: "That poor girl."

I was afraid to ask what she meant by that. I didn't really want to know.

The doctor told me that since I was now the proud owner of an irreversible cataract and three pieces of inoperable shrapnel, they would have to request my retirement. They would recommend that I appear before a board as soon as I had healed.

So I would be retired with pay and benefits. That sounded good. But, it wouldn't be easy to get used to civilian life all of a sudden. It was less than a year since I got my wings and bars. Now I found myself in a sort of limbo, a nowhere place.

I had become acquainted, at the eye clinic, with a young Ensign, another patient who came in for treatment. We'd traded sea stories. I asked him what ward he was in.

"Oh, I live off base," was his reply. "I'm married."

"You can do that?" I asked.

"Sure. I just report in here once a week. The rest of the time we go to the beach, take short jaunts around the countryside."

"Hey, that seems really nice."

"We enjoy it. Come over for lunch sometime, meet the wife."

"That sounds great. I will."

Hmm, I'm starting to get ideas. Here's something better than hanging around the sick and wounded in the hospital atmosphere.

All I have to do is get married and rent a cute cottage somewhere and wait for my retirement.

By the end of May, troops in Europe had marched into Rome and were poised for a landing in France. In the Central Pacific the Marines had taken Saipan, and continued to island hop as they neared Japan.

My left eyelashes were sheared. I had new battle ribbons over my left breast pocket, under my wings on a jacket that draped over my left shoulder hiding the cast on my arm; the empty left sleeve flopped around. We were sitting at a table, Roz and I, having our picture taken on our first return visit to The Coconut Grove since our engagement in December. We were back, celebrating my survival. There could not be a more poignant way to rejoice than there in that place where we fell in love. And Freddy was there too. The same old sentimental tunes reverberated through our bodies and souls. I knew I was home, safe at last. The music told me so.

Strange—as I wound up flight training the year before, small-town-guy that I was, I thought of home as the place of my roots, in Oregon with family and old acquaintances. But there I was in Glamour Town, acting as if I had grown up there, feeling like I belonged.

The place was loaded with men in uniform, a lot of us with battle ribbons. Our lives were accelerating, Roz and I. Five months before, we had celebrated our engagement there. I knew nothing of combat as it really was. I could now say unequivocally, I was a combat survivor who had a sea story to tell, with a love story in bloom as the music accentuated the action.

What a fortuitous moment for a proposal. Of course the ring is already on her finger but now I can propose that we actually tie the knot and find the cute cottage whose fantasy beguiles me so. I told her about my new friend in San Diego who lived with his wife in a cottage on a hillside overlooking the harbor: Naval vessels, two airfields, the civic center, aircraft factories... a busy hub of wartime activity.

"So we could do this too," I said with restrained enthusiasm.

"But we could find a place closer to the beach... a quieter place."

"We? Don't you have to be in the hospital?"

"No, not anymore. They've done all they can. I'm just waiting for retirement orders now. We could wait together. I could find a cute cottage for us. What do you say?"

She looked me in the eye, formed a small smirk with her mouth, eyes wide, "Ooh. This is so sudden, too sudden. I'd have to lose weight. I'm not going to stuff myself into a wedding dress, and... and I just started a new job."

"You look just right to me," I tell her.

Freddy was playing "DELIGHTFUL, DELICIOUS, DELOVELY", just egging me on. His timing was perfect, but I wondered about mine.

"I want to take you home, darling. Our home."

She just didn't have the same enthusiasm for the idea as I. Her two reasons sounded more like excuses to me. There may have been a little, but real, underlying, unspoken reason but I didn't want to push it, so I just sat back and enjoyed the music, the ambience... *Looks like I'll just have to renew my courting efforts. This is not a time to interrupt this memorable celebration.*

"I'll be going to my new home in the country this week," I told her.

"What do you mean?"

"Well, there is a special home, a sort of annex, loaned to the Navy for the duration, to be used by hospital patients who are on the mend or waiting for orders... a sort of convalescent home for officers. It's called the Burnham Home."

"Did you say it's in the country?"

"Yeah, in Rancho Santa Fe, out near Del Mar."

"Hey, pretty ritzy. That's an elite community. Mary Pickford and Douglas Fairbanks used to live there. When are you moving?"

The band was playing "YOU OUGHTTA BE IN PICTURES".

"Tomorrow," I replied, reluctantly, "I'd rather be moving in with you. To our place."

"We will, Darling. Don't worry. We will. When we're ready."

Oh, I'm ready, I said to myself.

Freddy and His Men were playing something familiar, not our song but one of his usual often-played tunes: "TIME ON MY HANDS" followed by "EVERY DAY'S A HOLIDAY". *Is he reading my mind?*

The Pilot meets his Dreamgirl

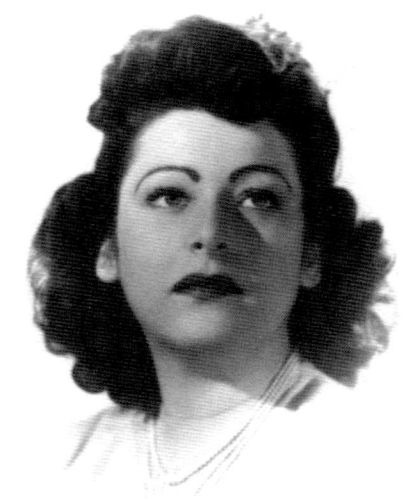

They go out dating in "Little Willie"

She wears his wings

He soars

He pitches

He flips over her

And they fall in love!

*Her picture is
in his wallet
His ring is on
her finger
When he goes
overseas*

With the Flyin' Lion Squadron and their 'Dreamplane"

In combat, his left eye and left arm are injured, he has malaria, is forced to come home.

Roz and Louise go see him in San Diego and at Del Mar Beach

While he heals, she visits him at the Burnham Home, the U.S. Naval Hospital Annex at Rancho Santa Fe, California.

She helps hasten his healing

They celebrate at the Coconut Grove in L.A.'s Ambassador Hotel

... the cast is off

. . . and they cast off on the swirling
waters of Holy Matrimony

And honeymoon at the Samarkand in Santa Barbara

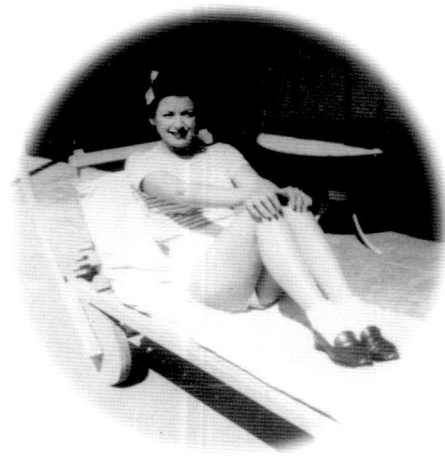

They dream of going to Stanford

They leave their wonderful home with Nellis and head Northward.

In the Bay Area, he reminisces about his days as a cadet there, in Golden Gate Park . . .

and his first liberty with comrades at Berkeley's Claremont Hotel.

They second honeymoon at Timberline Lodge

They settle in Stanford Village where Roz is caught napping after a hard day at the Veteran's Administration Hospital.

PART II

CHAPTER 9

BURNHAM HOME CAPERS

I was at the annex the next day, settling in to that grand two-story hacienda with its gated inner courtyard and balconies above. My upstairs bunkroom had a view of eucalyptus covered hills and vales with glimpses of neighboring estates.

About 40 patients lived there. I introduced myself to some of those in the sunken living room who were gathered around game tables and a baby grand piano. To one side, in an alcove, some older men hovered around a console radio. Through a large window I could see others on a neat putting green just outside and more men in swim trunks on the pool deck.

The occupants of that handsome estate varied in rank and stages of recuperation. They ranged from Aviation Cadets, who apparently came in for observation, to a Lt. Colonel, Marines, who was awaiting orders. Some were still healing or garnering well-deserved rest and recreation.

We even had a small chapel in a quiet corner below the main floor. We were under the care of a nurse and a cadre of hospital corpsmen. Our commander was a doctor who lived in the Village with his wife. That sort of thing seemed to be going around.

Our ambulance-taxi would take us to the golf course in Rancho Santa Fe and to the race track or the beach in Del Mar, where the "surf meets the turf." There were regular runs to the hospital during the day, and regular runs to the cocktail lounge in the Village or the bar in Del Mar in the evenings. We had just about everything anyone could desire right there at hand in that location, location, location.

So, how could I account for the empty spot I felt? Something was missing. The image of that cute cottage kept popping up and tantalizing me. Well, it was certainly no place for self-pity and I'd just have to settle in and enjoy it all.

After a few days I began to catch on to the mood of the place, a kind of laissez faire, loose with the "regs" attitude; amusing capers of some of the characters came to my attention almost every day, and every day was a holiday. According to "regs" or policy, we were allotted a three-day pass any time we wanted. We could take them in succession, but no more than three in a row.

One Navy Lieutenant passed his time traveling. He found a simple way to stretch the three days to a week or more. The Skipper, who lived in the Village with his wife, came in once a day to sign passes and requisitions. When he signed, he simply made a big stack and whipped through it with quick scrawls at the bottom of each sheet, never looking at the sheet itself. My friend the traveler left the days and dates blank and filled it in later when he estimated how long it would take to get to New York or Florida or whatever destination he had picked for the week. He caught military flights from among the many nearby bases.

A youthful Naval aviator, a Lieutenant JG who flew missions over Rabaul, a freckle faced kid, really laid-back, wore a T-shirt most of the time. We called him "Hot Shot Charlie" because he was a ringer for a character from a Milton Caniff comic strip of the day, *Terry and the Pirates*.

One day we were waiting in line outside the dining room when Hot Shot came strolling up in white T-shirt and khaki pants. An unsuspecting newcomer to our group, a Lieutenant Commander, eyed him up and down with an uncommonly stern expression.

"Don't you have something a little less casual to wear in the dining hall?" he asked.

Charlie, who probably hadn't worn his uniform for over a week, said, "Hell, it's only lunch."

But the Commander persisted. Hot Shot thought he was kidding at first; until he read the solemn look, then a smile came to his face as he turned away and hastened down the hall to his room. The smile was still there when he returned a few minutes later sporting a green T-shirt. It was contagious—we were all grinning—all but the Commander. He self-consciously gazed at the ceiling. He must have realized rank was mildly tolerated in that place.

The lowest in rank were the Aviation Cadets. I never found out why they were there. Recovering from wounds? For observation? Waiting for orders? On the day one accused the other of stealing his toothpaste—claimed he squeezed it from one tube to the other—I suspected paranoia.

But the most interesting guy, I think, hailed from Hollywood by way of submarine duty. He also was under observation, but not dangerous—far from it. He was there because he turned butterflies loose in a sub, an act so sane it immediately brought him under suspicion. On his first tour of duty, two or three days out of Pearl and headed into deep blue Pacific waters, his butterflies hatched out. He was so enchanted he wanted to share them with shipmates. They (the butterflies) were soon flitting about in a cabin below the surface of the sea. When the skipper heard about it, he immediately turned the boat around, headed for port, and had the young guy committed for observation.

We found out that Bob, the sub guy, had been a scriptwriter for a studio in Hollywood. He had a vivid imagination and a recorder, the kind with a wax disc. He was probably recording ideas for just the kind of scenarios that happened around the Burnham Home almost every day. One day a bunch of us were in his room looking at the recorder when someone came up with the idea of making a fake radio news broadcast. We could play it in the alcove where the radio console sat. We started to focus on a fake news broadcast about a disastrous event that took place overnight in the Philippine Sea:

> This morning before dawn a Japanese torpedo struck the Enterprise amidship at the conning tower. Admiral Halsey was standing at the rail in that location. It gave way and the Admiral fell overboard along with other high-ranking officers. They were immediately picked up by an enemy torpedo boat that darted in and out before anyone could stop it. The Admiral is suddenly a prisoner along with an undisclosed number of senior officers. It is not known if the Japanese actually planned to capture Admiral Halsey or if he was just the victim of fateful circumstances. We now have news from a Navy spokesman with the latest word on the disaster that occurred early this morning in the Philippine Sea... please standby.

A different voice came in with sound effects mimicking a short wave radio—a distant voice as if coming across the Pacific...

> It is true. Admiral Halsey is now in the hands of the Japanese Navy. All our search planes have already taken off and are in an intense pursuit of the torpedo boat that captured him and his offi... will... eep you... formed...
>
> Ladies and gentlemen, the overseas broadcast you just heard has faded out. Stay tuned here and we'll bring you the latest news on our Navy's effort to rescue Admiral Halsey...

When the recording was finished, we worked out a plan to play it for our unsuspecting colleagues. Our targets were actually the older guys who had been huddling around the console almost every news hour.

During the night, Bob would slip into the alcove, unplug the console, then plug in his recorder-player. Next morning, when the older gents started drifting toward the alcove, Bob would move subtly in and when they discovered the radio didn't come on, he'd say, "Oh, it's not working. I brought my table model in. Here—I'll turn it on."

The needle was all set so when the disc started spinning with the hatch down, the hatched-up broadcast began. The news hounds were soon hooked, their mouths dropped open. They looked at each other and said, "Oh my God... did you hear that? I'll be damned. They got old Halsey. I can't believe this!"

Some of them started for the main room to spread the news so Bob lifted the lid, moved the needle back to the start, and played it again. This time he left the lid open. It was time to show our joker. As others came closer to hear the "news," they could actually see the little yellow disc slowly spin its deception before their very eyes, and they were still buying it.

"Hey this could get out of hand," said Bob as the broadcast neared the end. He reached in and lifted the needle. Reality struck home with the sudden interruption.

"Hey, what is this? We bin duped. We bin suckered. It's a phony recording!"

We busied ourselves getting the recorder out of the room, the console plugged back in. Any new news would be good news now.

It was only a few days later when allied troops started the invasion of France: D-day on the beaches of Normandy. This was good news. It meant the war was definitely at its turning point, but not without loss and sacrifice.

This was the day, I learned later, when Charles Lindbergh wrote in his wartime journal:

Heavy tropical rain and wind during the night blew through the screen sides of the tent and dripped down through the roof. I covered my belongings with a blanket and moved my cot around until I found a reasonably dry corner.

First part of the afternoon writing, then I went to VMF 218 readyroom for a patrol over Rabaul. We took off at 2:00. Four Corsairs. We flew across New Ireland directly to Rabaul and circled the area around the city and its volcanoes. Dropped down to 3000 feet along the north coast and made one strafing run. Then over to Duke of York for several strafing runs, all ending within one hundred feet of the treetops.

Major (Ian) McNab and the other pilots sight a group of boats on the water, "almost as big as barges." McNab orders a line abreast strafing run. I have not seen any boats, but pull up to my place in line. Around a cloud and dive down toward the middle of the bay. There are fishnets in the middle of the bay that look a little like boats. A burst from McNab's guns, bullets spatter a hundred yards short. The next burst is on target. I fire a burst, wondering what it's all about. I drop down to within fifty feet of the water. Yes, they are fishnets!!! We join formation again, climb and fly along the west coast of New Ireland for a quarter hour. Then our patrol is over. Four more Corsairs arrive to relieve us, and we head back for Green Island at 4:40.

A message comes over the radio that the invasion of Europe has started; the Allies have landed on the north coast of France! The news spreads over the island like wildfire. I go out into the night and look up into the sky and stars and moon. The invasion of Europe! It is impossible to realize it on this island in the South Seas. I cannot project myself halfway around the world and realize what is happening on that "north coast of France." The destiny of the world is being shaped in the battle which is now raging—all hell broken loose along the English Channel. Millions of men attacking-defending-dying. And it is so peaceful here tonight— palm trees silhouetted against the moon-white clouds sharing the rest of the sky, the evening cool for the tropics, the camp quiet.

"The Lone Eagle" flew with my squadron on D Day. It was another "Lindy Hop".

Roz came to visit me at my country estate. There were no guest rooms. She stayed at the Del Mar Hotel with Louise who had just received the dreaded telegram from the Department of the Army. Her husband was killed in Europe. Roz was helping her through it.

I showed Roz around, introduced her to my new friends on the putting green, at the pool, took her to the beach in the bulky olive drab limo with the big red cross on a white circle. I couldn't swim yet—still had the cast on my arm.

The next weekend I resumed my commuting practices, another hundred-mile jaunt to L.A. but now, without a car, I took the train, the El Capitan. Roz was staying close to Louise, helping her through her personal crisis. They picked me up at Union Station and we all went out together, a somber threesome attending shows, dining. I stayed with the Whites. I felt awkward without a car, depending on others for transportation. *And our relationship— it's like courting all over again.*

She is thoughtful and wise. She is right to think about the future. I was probably right, when I first came back, to be cautious about holding her to the engagement... but then the nearness of her brought it all back and I could only think about getting a quick hitch up and a place to live together. But then it got even worse.

One night we were on our way home from an evening out in Louise's car; it must have been the week after my cast came off. I was driving. Roz sat in the middle, Louise to her right.

Roz said suddenly, "Oops, can you pull over to the curb and

park? My ring just slipped off my finger."

She'd been dieting and her ring had gotten loose since she lost weight. She had been in the habit of fiddling with it, sort of spinning it around. *Is there something significant about fiddling with engagement rings?*

I pulled over and we all climbed out carefully and searched around in the dark, in the seat, on the floor, in the street, in the gutter, but we didn't find it.

We drove to the nearest gas station where the light would make searching easier. But this proved fruitless and we gave up for the evening. I told them I'd come tomorrow and really go over the car's interior.

The next morning I went to Louise's and borrowed her key. I was convinced I would find that ring. I opened all the doors to give me full access and the best possible light. I took off the seat covers, looked in all creases, crevices, gaps, floor mats, slits, flaps, and spaces under, in, and about the seat, but found nothing. It just wasn't there. Our engagement ring was gone. It might have fallen out on dark Western Avenue when we got out to look. I even took a bus out there the next day and searched the place we had stopped at. It just wasn't anywhere to be found.

I felt a deep sense of loss, not just for the ring. Our relationship wasn't meeting my expectations. Roz had been losing weight and that contributed to the loss of the ring and there just wasn't as much to hug anymore, either.

In July, I noticed a tiny lump had grown on the now-healed scar beneath the lower lid of my left eye. It grew daily as my golf score improved—a game I had started to play in my teens. Now I had returned to this pastime because of the great golf course in the village and my need for an exercise that wouldn't endanger my eye.

I kept close watch on the bulge on the scar. It seemed to grow as the weeks passed. Then one day I noticed a dark spot had appeared in the middle of the lump just as my golf score approached ninety. Then, a tiny speck began to expand. The morning I broke ninety it had become so uncomfortable I decided to share what had popped up with the medics. I didn't even wait until morning. I took the noon ambu-bus into the Main Hospital.

Since I was not on the appointment list, I had to wait. Time passed and I finally got a chance to share my find with one of the ophthalmologists. He seemed excited—wanted to open it up at once to see what it was. He seemed sure it was the little fragment. I lay back in the reclining chair while he instilled a topical anesthetic. The surface was soon deadened. I held the lid down. He made a short incision with a tiny scalpel and voila: out popped a small metal chip the size of half a match head. Remarkable! I wondered at the fact the piece worked its way out through the original incision. Sort of a continuous running theme about this whole scenario: my last second escape from the cockpit of a disabled aircraft; working my way out of the jungle to a river, out the mouth of the river at night, out of the one man raft to the safety of a rescue seaplane. Everything seemed to have a way of working out... right down to that little fragment. I thought, *this bodes well for my relationship with Roz. Somehow it will all work out.*

A nurse put a big bandage on my eye. I thanked everyone for the successful treatment and hurried out to catch the ambulance. I looked around. It was not in sight. I looked at my watch. *Oh, the last one for the day has left. I could stay here for the night... no toothbrush. I really want to get back to my own bunk. I could go out there on the boulevard and hitch a ride... but it seems kinda funny to be hitchhiking right after minor surgery... with a big ol' white bandage on my eye... aw, I don't care. My legs are in good shape... nothing wrong with my thumb.*

I had only to hitch a ride to Del Mar. I could catch the early evening ambulance from the hotel. I was wearing my summer khakis with my wings and ribbons... *should get a ride okay.* And I did.

In the ambulance at the hotel I met a new patient. We introduced ourselves. It was the beginning of a long friendship.

His name was Stu... a young Ensign whose home was Los Angeles. He asked about the bandage on my eye. I told him about the "big bang" and the shrapnel and the surgery, how it worked its way out through the scar. He was impressed with the story and when I told him how I missed the ambulance and had to hitchhike, he said, "Hey, you're 'Fightin' Jack', aren't you?"

From that day on, as far as he was concerned, I was "Fightin' Jack." He called me that all the time.

Stu was the perennial sophomore, always on the fringe of excitement, always, looking for fun and good times. I don't think he'd seen combat. He was in the hospital for minor complaints, awaiting orders. He was a fraternity man at the University of Southern California, played on the baseball team. He was a good guy to have around for spirited companionship.

About this time Dave, Stu's antithesis, a Marine infantryman and casualty of the Battle of Saipan, limped into the Burnham Home. I knew him before our overseas mishaps. He was a member of the small cadre of service men who benefited from Bob and Nellis's hospitality in L.A. We had met a few times in '43. He had a serious hip wound from his part in the amphibious landing in the Mariannas. Dave showed a serious nature, had a harrowing experience in combat: a no nonsense kind of guy. He and Stu had little in common.

It turned out that I became the catalyst that formed this unlikely threesome: Stu, Dave, and Fightin' Jack. I went to visit Roz at her new place of work at a dental office in South Central L.A. one Friday in August after a train ride, a streetcar, and a bus—a trek that left me hot, tired, and sweaty. *This is no way for an ex-flying officer to get around,* I thought. *While I wait for Roz to finish work I'll mosey on down the street to that used car lot I saw.*

A couple of hours later I sat in my "new" 1940 Studebaker coupe in the dentist's parking lot ready to pick up my sweetie for a relaxing drive home. So, we had a different kind of weekend. I stayed with the Whites. I told them I would bring Dave up the next week in my new chariot.

It so happened that Stu's parents lived about a mile from the Whites and he jumped at the invitation to ride with Dave and me. Thus was formed that unlikely threesome.

We could try out the gas stamp gambit my friend "the traveler" had discovered. Not only would the Skipper's careless signing of passes net us each "leaves" of our choice, but we could take them to the gas ration board and get one gallon for each day of leave. We tried not to be greedy on our first attempt so 21 gallons would do us well for a weekend.

I'd been urging Roz to set a date. I guess the original courting period left something to be desired; now I felt as if I was starting all over again. She still insisted she had to lose weight. *Is this true or is she stalling? Maybe if I had a job, more money. At least I'd feel better.*

I had been playing more golf, getting better at it. I broke ninety recently and my score was getting lower. What a way to fight a war. I was not that gung ho about golf but I loved the links... a great place for a walk. Then I heard they were having trouble keeping the fairways mowed and the greens manicured. The course *was* looking a little scraggly. Not up to the standards when Sam Snead, Bob Hope, and Bing Crosby played there. Workers were hard to find: too many employed in defense plants or in agriculture.

The management had offered jobs to some of the guys at the Burnham Home. It was ideal: we lived in the neighborhood, we were in relatively good shape, and we could use the outdoor exercise as well as the extra pocket money. It may not be legal, but would they court martial a bearer of the Purple Heart? So, we enlisted, three or four of us.

One of the guys was a torpedo bomber pilot who had been in for mental fatigue, under observation for strange behavior or something (nothing as serious as turning butterflies loose in a submarine). Our first day out he flew the pickup we were riding. We zoomed across the fairway, the bomber pilot at the wheel. He suddenly shouted, "Look out!" and jammed on the brakes, but too late for a bunker—a drop off that sent us sailing through the air. We landed with a rattle, bang, kerchunk! A bone-jarring thud! I blinked my eyes, looking for signs of a retinal detachment. The sudden stop had shaken my bones but my eyes were okay.

We got out and surveyed the damage to the truck. Both front wheels were askew. The crash stretched the tie rods and left the wheels akimbo. The pilot drove the truck slowly to the equipment yard. He spent the remainder of the day realigning the wheels. The rest of us whacked our way through the area between fairways, cutting brush, grass, and weeds, sort of smoothing out the "rough."

The next day, on a test run, the pilot took a shortcut across a too-narrow footbridge; wheels straddled the walkway, which bent the tie rods again and cambered the wheels the other way. He spent the rest of the day realigning the wheels again.

In spite of our initial foibles we soon made up a reasonably efficient team of greens keepers. We manned big mowing machines without incident, ran fine cutters over delicate greens, dressed the greens with fertilizer, and hauled fresh sand into traps and bunkers. Some mornings we had to start early to get sprinklers going on fairways long before playtime.

So now I had a job. With purpose in life, my morale soared. Life was good out there in the fresh air and sunlight. I was getting back in shape and saving money for a possible marriage.

A new form of recreation had come to pass. Our private limo with the red cross on a white circle took us into the Village Inn at Rancho Santa Fe, a great gathering place in the lounge where we sat and drank and sang. We had a song leader with an Irish tenor voice and a great repertoire. I spent many an evening there with my friends Dave and Stu. Every week or so, a new guy joined us. People came and left. I was one of the few constants. It was mid-autumn and I had been there since July. My friends insisted I tell my survival tale of the South Pacific every time a new patient arrived.

My doctors decided that they didn't have to retire me right

away, since the interocular foreign body that did all the damage was gone. I could go on what they called "limited duty." Someone would find a non-flying job for me at one of the Marine Corps Air Stations: Mojave, Santa Barbara, El Toro, or El Centro. I had a chance to request a place I'd like.

In the fall of 1944 I worked daily on the fairways and spent evenings at the Village Inn while our troops in Western Europe were liberating the low countries of the Netherlands, Belgium, Denmark, and France. Our Pacific forces were moving up the Island chains into mid-Pacific, retaking the Philippines.

Songfests at The Inn were much alike. We gathered shortly after dinner, picked up a beer at the bar, and found a seat in the lounge. As more of our people arrived we found ourselves rearranging the furniture so that the entire room was soon ringed with men singing robust marching and drinking songs. We sat there all evening. We talked. The entire lounge buzzed with sea stories and repartee. Suddenly our tenor broke into song again and soon the room was full of singing voices and laughter.

One night we had surprise visitors. Some of my friends who were not working at the golf course were free to pursue field trips. They found a production company doing a movie at the blimp base down in the valley near the racetrack. They were shooting a film called "This Man's Navy" starring Wallace Beery. Some of the guys struck up conversation with director William Wellman and some of the actors, telling them about our nightly songfests and invited them over to sing along with us.

It was a memorable evening. Wellman told of some of his exploits in air warfare during the first World War. Their motto was "Lafayette we are here," a reference to a payback from our Revolutionary War. They were a squadron of Americans flying under French colors early in the war.

My friends insisted that I tell my story. And I did. It was story time. The actors, Jimmy Gleason and George Chandler, had tales to tell, too. The latter called his, "Chic Sale" stories, mostly about outhouses. Chic Sale told so many outhouse jokes that they named them after him.

Wellman's wife was there too but I don't think she had any war stories and our nurse, Ruby, was there as well. Her war stories were about the Burnham Home, but she didn't tell them. Before Wellman left he invited us to come to Culver City and watch the shooting of "The Story Of GI Joe" starring Robert Michum.

It was election time for the nation. Roosevelt was running for his fourth term and I got to vote for the first time for a President. I was staying with the Whites on a weekend and Roz had come for dinner along with another couple. Bob was in the den tuning in the radio for a political speech by the president. Because I supported Roosevelt I stood by in anticipation. Then, I noticed that the tables had been set and I could smell hot food. It turned out that Nellis was a lifelong Republican and could care less about the speech and announced dinner. Bob, frustrated, wanted to hear the president but was up against his wife's unfortunate timing. We began to seat ourselves, the speech had started, and Bob hadn't come out yet. Nellis called him. He bustled quickly out of the den, red faced, and said gruffly, "I'll take a plate of food into the den" and Nellis said, "You will not!"

This turned into a shouting match ending with a nasty invective by Bob. Nellis, mortified, told him he didn't ever have to come to the table again. He left the room and the next day moved all the way out. Our happy home away from home was now in tatters.

We liked them both. They had been so good to us. We tried to divide our loyalties. He had a small apartment nearby. She kept the house. We visited his bachelor pad a few times until he got a girlfriend. Then we drifted away, but gravitated toward Nellis, the one with extra bedrooms and a great facility with meals.

As winter approached with Roz still fighting her battle of the bulge, a real battle by the same name raged in Europe at the German-Belgium border. Roz moved on up in her career, from working with the dentist in South Central LA to the office of an oral surgeon in Beverly Hills. She had lost even more weight than six months ago when the ring slipped off her finger. I even helped her with her move towards independence by teaching her to drive. A fast learner, she mastered it quickly. I left my car with her for a week so she could practice solo, the only way to learn fast, except it was also risky, without a license.

A Beverly Hills cop pulled her over for a faulty brake light and had to write her up for driving without a license, the only traffic ticket she ever got in forty-five years of driving. I had the light fixed and helped her get her license before she had to appear. She had only to show proof that she had the two problems taken care of and there was only a small fine.

January passed uneventfully. One other guy besides me had been at the Burnham Home since July. A complete turnover in patients had taken place since then. The stalemate in Europe had broken and the march towards Berlin begun. I still awaited orders to limited duty. I had sent a request for two duty stations: the Marine Corps Air Station at Santa Barbara or St. Mary's Preflight School at Moraga. I dreamed of teaching the survival course there.

Finally, at the end of February my orders came: to report to the Marine Corps Air Station, El Centro, California. What a disappointment that was. Now I would be even farther from L.A. than I had been so far, and it would be in the Imperial Valley where it got hot and humid in the summer. Now it'd be even harder to get Roz to set a date. The thought of looking for housing in that hellhole was almost as bad as having to check into a BOQ and dine at Officers Mess. Breaking this news would not be easy.

CHAPTER 10

AN ULTIMATUM OF SORTS

I called Roz to break the news. She sounded noncommittal, not very excited, but congratulatory. She said, "I'm glad you're well enough to get back to work."

"Yeah, well I've checked out of the hospital. I have a thirty-day leave and have to report for duty the last week in March. I'll pack up my stuff and drive to L.A. Please think about a date to get hitched soon so there will be enough time for a honeymoon." No answer.

I didn't know what to tell her about El Centro. I had never been there but I heard it sizzled in the summer, that it had a reputation related somehow to the French Foreign Legion. "I'm sure we can find a place to live." I cautiously tried to reassure her that we would do fine, all would be well. She showed an air remarkably neutral. I had no idea what her innermost thoughts were.

"We'll see," she said.

That sounded like my mom when I was a boy.

I got most of my gear stowed in a footlocker and a suitcase and tossed a few papers, pictures, and letters in a small box. All these, all my earthly belongings, went into the trunk of my car and I was off for L.A. with one thought in mind: to talk Roz into an early marriage.

I stayed in one of Nellis's spare rooms. Roz continued on her job. I drove her to work every morning and took her to dinner most evenings, doing my best to convince her that she'd blow away if she got any thinner; that her old excuse was useless now. But my efforts seemed futile. It looked like I was destined for bachelorhood on the base, life in a BOQ, food at the Officer's Mess, drink and entertainment at the Officers Club.

Finally I made what amounted to an ultimatum of sorts. Part of my leave had flown by already. I wanted to go home, visit my family and friends in Oregon.

"I will catch a flight to Portland right away," I told Roz, "call me any time. Try to set a date before my leave is up. I'll be with my parents until then... unless you call."

I told her I'd leave my car with her. She could use it to drive to work. If she didn't call I'd pick it up the day before my leave was up. That's the best I could do. *Who knows what happens after that?*

She drove me to the Naval Air Station at Terminal Island. On the way I thought, *This could be a turning point in my life, the start of a separation. At the airfield, it won't be a last farewell, but a poignant moment approaches.* When we finally arrived at the field, a scenario started to unfold. I was reminded of scenes in movies, dramatic farewells.

I thought, *What was that line in Casablanca? This could be the beginning of a beautiful friendship? Naah, Rick was talking to Louie.*

I can have Sam play As Time Goes By. That's your favorite tune isn't it?

Naah, our song is People Will Say We're In Love.

Let's see, We'll Always Have Paris.

Naah, we haven't been there yet. But I'll think of something.

We stopped near the loading ramp. I climbed out, grabbed my bag, went around the car, leaned in and gave her a big kiss then opened my mouth, about to say, "Here's looking at you kid." But I stifled it with a bland but urgent, "I hope you call."

My parents were happy to see me—to have me around for a couple of weeks. I felt kind of strange, as if I had been cut loose. Relationships become part of you and when you verge on separation, it feels almost like a pending amputation. Being back in familiar territory helped ease the discomfort and I was soon buzzing around Portland on visits with relatives, young cousins who grew up suddenly, old friends whose lives had been changed by the war.

Soon I had almost forgotten the ultimatum, when, one evening a phone call caught me by surprise. It was Roz She had decided. She'd set a date: March 16.

I said, "This is so sudden. What made you decide?"

"I'll tell you later," she said. "Can you be here Saturday? I made an appointment for pictures."

"I sure can," was my eager reply, "I'll be there in a flash for the flash."

So, now I had to suddenly break off the visit with my parents. They said they wouldn't be coming to the wedding. I apologized for the quick change in plans and the sudden announcement of a wedding.

Dad said, "If you love her you should marry. I'm for that."

Mom was quiet. She was not pleased, not happy about this turn of events. They met Roz once in Los Angeles when our romance had only started. She helped them get a hotel room and met them at the train station on their first trip to the southland. The signal was a rose on her lapel. But Mom didn't expect her exotic looks. She thought I would find someone with more of a "small town look" and it took some time for them to approach her.

Mom's explanation was pretty blunt, "We saw the rose but you didn't look like someone Jack would go with." Well, anyhow, I had Dad's blessing and that meant a lot to me.

Back in L.A., I was ecstatic. On the way to the photographer's I asked her again what made her decide so quickly.

"Wait 'til we get there."

'Why?"

"I have to show you."

When we parked she asked me to open the trunk.

'What for?" I asked.

"So I can show you why I decided."

"Oh."

I unlocked the trunk and lifted the lid. She pointed to the box. I asked, "What?" And she said, "You know. Those letters from girls with Oregon addresses and you were loose in Oregon. So I decided."

We entered the building—with me wiping away an embarrassed grin—in preparation for photographs at a well-known Beverly Hills studio. Our pictures would be in the Los Angeles Times. I had been too cheap to buy formal dress blues so I wore my summer gabardine khakis. Roz was beautiful in a gray suit and a bonnet piled high with bright little artificial flowers. I sported a row of colorful ribbons under my Purple Heart ribbon and my gold wings over my left breast pocket. Her smile revealed a row of pearly teeth. We were a pretty couple.

I stayed with Stu at his parents' home. On the day of the wedding we were doing gymnastics in the living room. Time slipped by and we were completely distracted by our activities. Suddenly Stu's mom reminded us that we should start to get ready for the wedding. "My god, I almost forgot. I'm getting married. Time to shower, shampoo, shave, don a clean shirt, shine my shoes. I'm getting hitched."

Roz had chosen the Chapman Park Oratorio as the site for the ceremony. It lies across Wilshire Boulevard from the Ambassador Hotel and the Coconut Grove, very close to the Brown Derby and the Rendezvous Room. The Oratorio is a small chapel on the grounds of the Chapman Park Hotel. A Navy Protestant Chaplain performed the rites. Rosalind's parents, her friends, "The Club," and relatives were there, along with Nellis and Stu's parents, to help us celebrate our union. There was no reception. We had some congratulatory moments. Then the four principals: June, maid of honor, Stu, best man, Roz, and I, walked across the street to the Coconut Grove where we celebrated to the music of Freddy Martin and his Men who played their hearts out to make it a complete evening for us, a sort of closure.

We didn't stay too late. Roz and I would be driving all the way to Santa Barbara that night for our own special form of celebration at the Samarkand hotel.

CHAPTER 11

MARRIED AT LAST

On the night of March 16, 1945 Rosalind and I set out on the road to a whole lifetime together; the beginning of the family I yearned for, to share with her, my beloved. The life would be, like the road we were driving now, rough and smooth, crooked, hilly, wobbly, through all kinds of weather and topography. We would survive all sorts of storms; enjoy all manner of pleasures, successes, and failures. We were on our way.

We arrived late at midnight and were soon in nirvana, in each other's arms at last.

Our week in Santa Barbara was mostly a relaxed and enjoyable mixture of sightseeing, swimming, sunning, visits to the mission, attendance at a theater with a starry ceiling on State Street where "Abie" and his "Irish Rose" gave us a good feeling about our mixed marriage with Irish husband and Jewish bride named Roz, instead.

At the end of the week we were on our way to El Centro in our Studebaker coupe, most of our belongings stuffed in the trunk. It occured to me that my Mom and Dad started their married life in the same make of car 27 years earlier.

We stayed overnight in the mountain village of Alpine and arrived at the Marine Corps Air Station, about six miles west of El Centro, in the morning. I reported in with my orders, met the operations staff and the base Executive Officer. They had been expecting me, wondered where I had been.

"My honeymoon," I told them, "I've been on leave."

"Oh, We didn't know. Welcome aboard."

I asked about housing. We checked in earlier at the Barbara Worth in town but it was too expensive to stay there very long. Major Waldie, the Exec, invited us to stay with him and his wife in their on-base quarters, one of only two houses on the entire base. The Skipper and his wife occupied the other.

We moved into their guest room the next day and Roz hung out with Mrs. Waldie while I started to learn the routine at Operations, part of Headquarters Squadron and the heart of the base—or maybe the brains. It was the communications center where all incoming and outgoing flights were accounted for, along with the crash crew, weather, parachute loft, and auxiliary aircraft. The latter was available to us for local flights at our discretion. Because of the condition of my left eye, I would not be able to fly solo. I would have to have another qualified aviator along on flights. But, I would draw flight pay.

A month's pay awaited me along with a promotion to First Lieutenant. I got a chance to log four hours of flight time with one of the other assistant operations officers just in time for my first paycheck. This, along with the promotion and dependent's pay, made me feel like a good provider at last.

While I worked in operations, Roz went out house hunting. My first assignment was a challenge. I was given a large flat plywood box, wired for lights inside. My job was to attach a huge map to the long side of the box. It would actually be a series of smaller navigation maps pieced together to become one large plot of the southern portion of California and part of Arizona, from Phoenix to San Diego to San Francisco.

I got all the pieces glued down nice and flat and lined up

squarely. Then I drilled holes in the spots where there was a major airfield—military or civilian. With the lights turned on, each of the fields showed up as a point of light. I made Lucite pegs to put in the holes to make them brighter. We changed the colors with extra pegs that had colored tips so that if a field had dangerous weather conditions it would show a yellow light; a red one denoted a closed field.

I attached a long cord at our base location and next to it a scale that could be used to measure the distances to nearby bases and airports.

This new navigational aid, fastened to the wall next to the operations desk, was available to pilots who came in to file flight plans. At a glance they could tell conditions at destinations and calculate distances.

Roz found a little cottage in the country about four miles from the base. It was a former foreman's house in the middle of a sugar beet field. A real sugar shack.

The first night there we had an encounter with a huge tarantula climbing up the wall of our bedroom. Roz screamed so loud she scared me more than the sight of the ugly creature. We gradually learned to deal with these denizens of the tropics. In the kitchen small scorpions frequented our utensil drawer. After carefully sliding the drawer open we'd grab a knife and quickly cut the first thing that moved into several pieces.

One morning I forgot to inspect a shirt before I put it on. Standing before the mirror as I buttoned, I saw a large dark form creeping out from the armpit. I jumped up and down in an attempt to dislodge it. Roz came in to see what was going on and cracked up as I danced around in a jig trying to shake the ugly thing. I finally managed to unbutton the top two buttons and wriggle carefully out of the shirt and shake off what turned out to be a vinegarroon. When I crushed it, it emitted a distinct odor of vinegar. It had a long body and two large antennae that gave it the appearance of a cross between a scorpion and a tarantula.

And so it went. We dealt with our creepy crawlers on through the spring as troops in Europe slipped the noose around Berlin. Hitler was finished, thank God, and we soon celebrated VE Day. The Okinawa Turkey Shoot in the Pacific was almost over and our flyers had more bases for bombing Tokyo.

The heat and humidity increased. You never dried off. The sweat came out and just sat there. We learned the El Centro salute: a swipe across the brow and downward flick of the fingers; ladies did the El Centro curtsy: after standing they bent their knees and pulled the back of their skirt loose with a flourish.

Some interesting episodes unfolded on my watch at operations. With the Skipper, the Exec, and the Operations Officer all off the base one day, I became the Operations Officer and the Skipper, the one in charge.

A flight of Navy fighters came in from San Diego for some pre-dusk exercises. I was in my office when the siren went off in the tower. One of the fighters had accidentally landed with its wheels up and our crash crew responded at once. I went out and got in the jeep that was provided for such occasions. I had a driver. On the way out to the runway I discovered that I also had a siren, which I promptly put to use for no particular purpose but to give notice that we were on our way. Upon our arrival at the crash scene we saw there was no fire, but the crash crew stood by with extinguishers at the ready. The plane's wings were right down on the runway so I easily climbed up to the cockpit where a young pilot sat embarrassed, head down, slowly recovering from the shock.

"Forgot... to... put... the... wheels... down," I said slowly, as he nodded.

I offered him a ride in my jeep as the crash crew brought up its wreckage removal equipment and his squadron circled above in the growing dusk like a pack of vultures. This young pilot would hear many mocking, unfriendly words about this.

CHAPTER 12

MY FLIGHT STATUS THREATENED

Shortly after VE Day a special order came for me from MARAIRWEST headquarters to appear at a board of aviation disposition. I was told the purpose of the board was to determine if I should continue my flight status, that is, if I would retain my eligibility for flight pay.

The board would convene in the administration building at El Toro Marine Corps Air Station. The procedure would take two days and involve another flight physical exam. The idea of the inquiry disturbed me. The thought of having my wings clipped concerned me. I loved flying and disliked the thought of losing flight pay. I'd have to stay overnight which meant Roz would be alone one night. She arranged for a kind of pajama party with some of the women marine officers she had befriended.

After landing at El Toro I was driven twenty miles to the Naval Hospital at Corona where a thorough examination of my eyes, as well as my other parts, was carried out. The next day I appeared before an impressive array of senior officers around a large table. Matter-of-factly, they pored over my medical records. I hoped they were able to read hardships and suffering between the lines of those coolly collected medical notations. They asked a few questions about my general health, then dismissed me without a hint of their possible decision.

Back in El Centro, I made a quick call to Roz to see if she missed me. After all, we hadn't been apart since our marriage. She told me how well the slumber party turned out and she missed me and she would pick me up right away.

A few days later the word from the board came about their decision. They'd decided I should be grounded. What a shock! I almost felt a sense of betrayal. I could still wear my gold wings but they would only be a nostalgic souvenir, a symbol of yesteryear. If the powers that be in Washington concurred with the board's finding, my flight pay would be cut off. I could only wait.

I drew the duty again by myself. This time the skipper was aboard. A call came in on the hot line from NAS, North Island, to tell me they were socked in and a Marine Corps transport on its way from Hawaii with a VIP aboard had decided to land at El Centro. They give me its ETA, which was less than thirty minutes away. I looked over the protocol checklist for VIP's and found I must first notify the base adjutant. He alerts all the necessary persons: the Skipper, color guards, base relations, photographer, car and driver, etc. The adjutant wanted to know who it was. I told him we didn't know yet. "It's a Marine Corps transport only minutes away. I'll call you the moment we find out."

"The VIP is probably a flag officer which means we'll have to hoist his flag the minute he comes aboard." The plane would have markings that show his rank. I alerted the people in the control tower to give the transport landing priority and to let us know the moment they spotted the telltale insignia. In minutes the plane entered the traffic circle. The greeting entourage waited expectantly on the tarmac; binoculars were trained on the plane's fuselage and tail fin, the adjutant anxiously awaited my call.

Then word came, "Four gold stars on a red field." It was the Corps' highest-ranking officer, General Vandergrift, Commandant of the Marines and a hero of the Guadalcanal campaign. He had recently been appointed Commandant and promoted to four-star rank.

I called the adjutant. "Hoist the four star flag," I said.

"Whaddya mean?"

"It's Vandergrift."

"Naah. Tell me who it really is."

"It's him. Don't you have a four star flag?'

"Of course I do; it just came in this week.'

"Lucky for you. You better get it up. He's on his way."

"Well, whaddya know. What a surprise!"

The plane taxied in and parked just outside the operations office. Through a closed window, feeling a certain pride for my part in the orchestration, I watched the whole ceremony in pantomime. The side hatch opened and out stepped this dignified gentleman with rows of ribbons arrayed across his breast and eight silver stars sparkling on his shoulders. He seemed delightfully surprised by the welcoming party. After all, he did just drop in unannounced. He smiled as he returned the Skipper's salute and was obviously pleased as he walked past the honor guard to his waiting car.

Finally, a letter arrived from Washington and with it an indication that somewhere in the Pentagon's vast compartments someone was looking out for my best interests. Powers-that-be had rejected the board's counsel to clip my wings. In fact, the awareness of some special watch over me began to awaken in my consciousness; the realization of a whole string of fortuitous happenstances which came on the heels of my own personal big bang; the loudest explosion I ever heard that sent me into the blackest of blacks; the worst thing that ever happened to me, followed by a whole string of happy, unaccountable circumstances. *What luck! Or is it?*

Perhaps there was a clue in the lingo we used in air combat, the code words we used such as "bandits" for enemy planes or "vector" for take-a-heading. The one that had my attention now was the word for each thousand feet of altitude: "Angels". After the big bang when my plane tumbled from Angels Nineteen to a place where it turned upside down at the last possible moment for me to unsnap the seat belt and fall out, it had dropped to only one—*one Angel left*—just enough to get out before my plane crashed into the jungles of New Britain. From that moment everything seemed to go my way in spite of some pretty stupid things I did. It seems as if that one Angel stuck with me from that time on, making it possible for me to find my way through a trackless jungle and to take care of my own wounds so the superficial ones were soon healed. The more serious wound was actually healed by a serious disease, malaria, which then burned itself out. I'm convinced my Angel protected me from enemy submarines and aircraft at sea, from enemy patrols on shore. *Why did the wind pick up and carry me out of dangerous waters into a safer place?*

Just as mysterious was the question of why someone in Washington reversed the findings of a cadre of high-ranking officers who had shown perfect rationality in conducting their inquiry. I could only conclude the one Angel still stood by. *What next?*

CHAPTER 13

WAR ENDED WITH A BIG BANG

The war ended... that's what. And suddenly, too. All at once we weren't doing anything for the war effort anymore. We didn't have to worry about loose lips sinking ships. Which reminded me. I recalled an incident at the Del Mar Bar when a civilian with loose lips, one "Happy Hour" in a lounge filled mostly with men in uniform, seemed far too willing to let us know he was no ordinary civilian. He started talking about a secret weapon he was working on up in Hanford, Washington. I turned and looked at him, because I remembered Hanford from my flight training at Pascoe when we saw unexplained bright lights on night flights off to the northwest

"Big explosives," I heard myself say.

I was only guessing, but his eyes widened and he looked at me as if to say, *I wish you hadn't said that.* He turned and quickly drifted out of sight—must have realized he'd said too much.

The war ended prematurely with two big bangs and I wasn't even ready. I might also say, I thought the aim was bad. We wiped out a lot of women, children, and other innocents when the sacrifice of a lofty peak such as Mount Fuji might have served as well to demonstrate the force of a weapon with a capacity for mass destruction. Our people who took aim on all those civilians just wanted to stop the war and save the lives of thousands of military combatants. But an embargo would take too long and maybe the loss of a mountaintop would be a waste of a good weapon?

The celebration at the Officers Club served as a wakeup call for Roz. At least that is when I began to notice a quizzical look on her pretty face. She was thinking about our future. She knew it was getting closer.

"I had better get back to L.A.," she said.

"Why?" I asked, stupidly.

"Because we are suddenly on the verge of gigantic changes. We can't just sit here and watch it happen all around us."

"Hmm, maybe you're right. I hadn't thought of that."

"I'll get a job and stay with Nellis."

We rushed off to call Nellis. She told us she'd love it and she was glad the war was over.

Within a week we had packed up and moved out of our "ranch shack" and into Nellis's home on Leeward Avenue in L.A. We felt lucky to have such a friend. Roz started her job search at once and I moved into a BOQ on the base, dined at officers' mess, and hung out at the Club playing pool and blackjack.

Roz got a job right away at a dentist office close to Nellis's in that high-rise art deco building at the corner of Western and Wilshire. We were back in a familiar routine. I now commuted farther than ever on my days off but I had aircraft at my disposal. I had only to cajole one of the pilots to chauffeur me to a field at or near Los Angeles. My return trips were usually anything I could find going to El Centro. Once, I hitched a ride in the belly of a torpedo bomber, sitting where the torpedo normally would. But the war was over and it was a free ride.

That was about it for the next three months. Then in November I was surprised with a set of orders that released me from active duty as of 1 December. This surprised me because I found no mention of the cataract in my left eye that left me with impaired vision. I went to the M.D. at the base sick bay and reminded him that I had been on limited duty because of my vision and asked him if I shouldn't have orders to a Navy Hospital instead. He agreed and proceeded to write up orders in lieu of the release from active duty. He even asked me what hospital I preferred and I told him, "Corona Naval Hospital." This would shorten my commute to L.A. by about 75 miles.

I had a chance to take one more flight before I checked out. I chose a Piper Cub and found a friend to go up with, someone who needed to log flight time. Lt. Bordigan was his name. We had flown together before. As we made our way to a nearby auxiliary field I remembered my first flight ever in exactly the same type of single-engine, high-wing monoplane, same yellow color. That was an orientation flight, my first time at the controls. This would be my last. A lot had happened in between. We shot a few landings and carried out some gentle wingovers above the field. We flew for about an hour then headed back for the main base where I shot my last landing, and with a catch in my throat, said goodbye to flying.

The Navy Hospital at Corona was really an old ornate hotel once known as the Norconian, a luxury resort and spa, converted to a hospital by the Navy. I was assigned a room with three others. One of my new roommates was really old. He had fought with Admiral Dewey in the Philippines and he liked telling about it. His tale was endless. It went from the battle of Manila Bay to a double date he had in Seattle, which reminded him of the war in the Philippines, and on and on it went.

I stayed there only long enough for the doctors to consider my case and make their recommendations. This took a couple of days; then they told me I could go home and report in, once a week. They had examined me thoroughly and concluded that nothing could be done for my eye, that the condition was irreversible. Two small particles remained, and the cataract. Now, I had only to wait once again. I would eventually appear before a Marine Corps Board of Retirement in San Diego. I could go home to my honey at Nellis's, where I practiced civilian life and bought some new tweeds and loafers—the collegiate look.

Date	Type of Machine	Number of Machine	Duration of Flight	Character of Flight	Pilot	PASSENGERS	REMARKS
November 1945							
2	NE	26355	3.8		OLSON H.J.	MORRIS J.G.	LOCAL / EL CENTRO
9	NE	26355	1.3		MORRIS J.G.	BURDIGAN	LOCAL / EL CENTRO

CHAPTER 14

ANOTHER WORKOUT

One evening in January, as I sat in my new tweeds and read, I noticed some discomfort in my left eye, on the inside of the lower lid, then a sharp pain adjacent to the scar, piercing as a pin prick. I had to hold the lid away from the eye to stop the pain. I went to a mirror and saw another dark spot in the middle of the scar. I called Roz and showed her my predicament. We tried to decide what to do. It was not a life-threatening situation but something had to be done. I couldn't hold my eyelid down all night. But what could we do? Then we remembered the Naval Reserve Armory. It was open twenty-four hours and they had medics on duty. We had only to drive down to Chavez Ravine.

Roz drove while I held onto my eyelid. At the armory we found a doctor and showed him the tiny little thing that was causing such consternation. He agreed it was tiny, yet sharp as a needle and needed to be removed. He instilled topical anesthetic in the eye, waited for it to take effect, then grasped the particle with forceps and out it came. My face seemed to have a propensity for rejecting foreign objects. Tiny pieces of metal and Plexiglas had periodically worked their way out of my nose and lips and now this, a second fragment had surfaced on my left eye. *What next?* As I once said, "everything seems to be working out."

Roz knew better than I about the need to plan for the future. I had difficulty in moving on to the next step until the present step was completed. She felt I needed to consider my education and preparation for post-war life: she saw the potential for degrees and training, while I seemed to be in a kind of stupor. I had all that time available, still drawing pay, waiting for word from Washington.

One day Roz saw an announcement in the paper about a Stanford entrance exam to be given at Los Angeles City College. She urged me to go and take it. At the library I found a recent Stanford catalog of courses. I wanted to see what they offered of interest to me. I noticed a course in the Art Department section called Industrial Design, a curriculum combining art and engineering. It sounded like something for me. I always wanted to take art classes but rejected the notion because my family thought them impractical. But this was now and art classes combined with engineering classes looked truly practical to me. You learned to design useful objects, products, make them interesting, artistic, then you provided a service and expressed yourself at the same time. This idea had my interest and I was going to pursue it. I would take that Stanford Exam.

I went in that day on a kind of a lark—relaxed, fun filled. I was not under pressure so I took the exam with a relaxed attitude, sort of devil-may-care. It all seemed easy.

Finally, in April, orders from Headquarters arrived. I was to appear before a retirement board in San Diego. I dressed up in my freshly pressed winter greens and polished cordovan oxfords and departed in my shiny 1940 Studebaker coupe for the Marine Corps Base near Lindbergh Field. It was only formality. Those senior officers on the Board would pretty much follow the recommendations of the medical officers. My retirement was based on a public law written for Naval Aviators who became disabled while flying in combat, to compensate them for the loss of their flying career. My retirement orders would be sent to Washington for the signatures of President Truman and James Forrestal, Secretary of the Navy, and until that time I would stand detached from active duty.

Right after that a card came in the mail from Stanford. It told me how I did on the test. I was placed in the top group, which I discovered later, meant the top ten percentile and that meant of all the people who had ever taken that exam. Students with such a score were sure to do excellent college work. I was surprised, delighted, charged, the lethargy faded rapidly. I showed the card to Roz. She lit up too.

"Hey," I said, "Lets go to Stanford. My retirement will come through this summer and we can enroll in fall courses."

She was excited. She had always wanted to go to Stanford, had even been accepted to go there out of high school but opted for Cal because her brother was paying the bills and it was less expensive. So now we would go together. We studied the catalog more thoroughly as we looked for class start dates and course descriptions.

By late spring we had decided that was what we would do and Roz gave notice at work for a May 1 departure. I could actually start classes in June. We had to find housing and we needed a head start because places to live were hard to come by everywhere.

There was one more piece of unfinished business we wanted to take care of before we left. Rosalind's mother had been going to a spiritualist for years and believed the woman had special powers.

She had a record of finding lost possessions. They had been discussing the possibility that she might be able to read the whereabouts of our lost engagement ring.

Roz conferred with a neighbor who had some issues that she wanted advice about. The two decided to try the spiritualist. Her office was in Hollywood. I went along just for the ride and waited in the foyer. Roz and Ann were told to write their questions on slips of paper, fold them, and place them in a basket sitting on a low table. There were a number of people, mostly women, seeking answers that night. I could see part of the room through a pair of open French doors but only murmuring sounds reached my ears.

On the way home they related what happened. She picked up the slips but she didn't unfold them. She just held them up to her forehead with closed eyes and accurately caught the meaning of each question. She not only answered all the questions but also added tidbits of unsolicited information.

She answered Roz' question but preceded it with, "You are very close to a young man who recently endured a frightening experience. He was lost."

Then she said, "You lost a ring; in or around a seat. It is still there."

She answered Ann's question and added, "You wear a ring that has a crack."

Ann checked the ring on the way out of the room and bingo. There it was: a crack she didn't even know about.

By the time we reached home Roz had already formulated a plan for searching Louise's car again.

CHAPTER 15

AND ANOTHER

I'll plan a shopping trip with Louise," said Roz. "She can drive her car over here. We'll go shopping in our car. That will give you time to take the seat apart and do a full search."

I agreed wholeheartedly and a plan formed in my mind at once. This time I would be more methodical, more thorough. Arrangements were made and Louise dropped by on Saturday to leave her car as planned. It turned out I got unexpected help from Nellis's brother-in-law who had dropped by with his wife for a visit.

I explained that Roz had lost her engagement ring in the front seat of Louise's car eighteen months ago, that we had searched the car twice without any luck, but were reassured by a clairvoyant that the ring was still there. Now I was about to make one last stab at a fully thorough search. We took the seat cover off and spread it out carefully on the lawn. Then we unbolted the seat, slid it out, and set it down on the seat cover. We turned it around, peering down into all the crevices, cracks, nooks and crannies, carefully searching every square inch of that seat with our fingers and we found... nothing. It was not there.

We looked at each other with disappointment and resignation written on our faces. I looked down and stared at the empty underside of the seat.

"Well, I guess we might as well put it back together," I said after another careful search through the floorboards... nothing. We picked up the seat, slid it back into position, and bolted it in place. I turned to the seat cover and reached down to pick it up when something shiny caught my eye. I focused on it—the ring! It sat there as if manifested, as if a mysterious hand had waited for the appropriate moment to place it in the center of the seat cover so it wouldn't be missed. It was one of those moments you never forget. For the rest of your life you see the sparkle of the found object, tantalizing your sense of reason. I reached down and grasped the prize, held it up to the light and blinked my eyes in disbelief. It blinked back. I wondered where it had been all those eighteen months, just waiting for a dramatic re-appearance.

When Roz arrived home with a question on her face, I told her to hold up her left hand.

The question turned into a bright, surprised exclamation, "You, you found it?!!" she cried, as she raised her left hand.

"Yes," I shouted triumphantly, and slipped the ring on her finger atop her wedding band.

"Where was it?"

"Oh, it was just sitting there, blinking at me."

I filled her in on the details, as much as I knew. The rest was a mystery. She was ecstatic as she thanked Louise for not selling her car, thanked me for being diligent. I told her a lot of gratitude must go to the fortuneteller. Things worked out...with a lot of help from your friends.

Since we had found the ring, it seemed all unfinished business had been taken care of. My retirement only awaited the President's signature, which would happen on 1 August. All we had to do was pack our stuff and head out for Stanford. We bid a fond, grateful, sad farewell to Nellis (she had been so supportive and generous to us) and her young son Bobby. He was only four when we started coming to their abode, now he walked off to school by himself. We could probably have rented a room somewhere else but there's nothing like being in a home where you feel like part of the family. Which reminded me of our destination: our first stop would be at the home of Roz' brother Milt, in Lafayette, California.

CHAPTER 16

NORTHWARD BOUND

Lafayette is a suburb near San Francisco, where Milt, my brother-in-law, lived with his wife, Helen, and their young toddler, Johnny. We'd be staying with them while we searched for housing down the San Francisco Peninsula and took care of business at the Stanford University admissions office.

Roz made a preliminary housing search on campus bulletin boards while I checked in at the office of the Registrar in the Administration building. I walked in and flashed my card, the one that placed me in the upper group, the result of the aptitude test I had taken at Los Angeles City College. They asked me if I had filled out an application.

"Whaddya mean?" I asked. "I thought that was what I did in Los Angeles."

"No, that was a college aptitude test," they explained. "You have to fill out an application and include high school and college transcripts."

"Can I do that here?" I asked.

"Sure, and fill out these requests for transcripts, too."

It took awhile to fill out all of the forms. When I finished I handed them to a clerk with a question, "When will I know? Summer quarter starts next month."

"Everything should be here in two weeks," was the answer. "and another week for review."

As I walked across the campus to my car, I thought, *I took a hell of a lot for granted.*

Roz was waiting. She had found an ad for a room in Palo Alto. As we drove away she asked me how it went.

I gazed across the Oval at the entrance to the University, with its Romanesque arches framing the mosaic mural on the facade of Memorial Church in the middle of the inner quad. "Boy, I really like this place. I hope they take me in."

"Why, what's the problem?"

"My application is not complete yet. They need my transcripts. We can find a place to live. Then we have time for a trip to Portland, visit my folks, take a ski trip to Timberline Lodge on Mount Hood."

My spirits were lifted at the thought.

"We'll find our nest and forget our problems for a couple of weeks."

Roz had never been in Oregon and I looked forward to showing her the land of my youth.

We drove in to Palo Alto to inspect the room she had located in a classified ad. It was in a large house on Fulton Street, not too far from the downtown area. The owners, a pair of attorneys with a small boy about three, had rented one of their rooms to another Stanford couple. There was still a room left with attached bath. We decided to take it. We could move in on the first of June. By then we would know if Stanford was our destiny.

Back in Lafayette Roz visited with Helen while she packed for the trip. Milt was about to get his release from duty at the Navy Pre-flight School at Saint Mary's College in Moraga, where he attained the rank of Lieutenant Commander while working as the officer in charge of the Ships Service. I went with him to the base for old time's sake. I had started my training as an aviation cadet there four years before. It would become a college again with a lot of new buildings and athletic facilities. It was one of the places that had contributed to my physical attributes, the strength and stamina that had in turn made a huge difference in my ability to survive the rigors of those challenges in the South Pacific.

As I sat and sipped an old fashioned at the Officers Club on the hill, I gazed out over the vacant athletic fields and parade grounds, the academic buildings and dormitories. Visions of all of those many and varied activities passed in review: bouncing high on the trampoline in the gym, dressed in full football regalia on the field in the middle of August, dashing for the finish line at the end of a 2-1/2 mile run on a hot afternoon, marching with a rifle in full uniform across the parade grounds, swimming and diving in the pool, hand-to-hand combat, and a challenging session on the obstacle course. Even a night of recreation in the gym came back.

Late in the summer of 1942, while our Navy had fought the battles of Midway and the Coral Sea and occupied Tulagi and Guadalcanal, we aviation cadets had endured three months of grueling physical and academic training. We were rewarded with a night out. Not all the way out, but to a smoker in the gym, where we rooted for our favorites on the wrestling and boxing teams. There seemed to be a lot of excitement in the air, a lot of built-up energy.

The Navy's Pre-flight Program was designed to bring the cadets up to their absolutely highest possible physical and mental potential. It had been designed from top to bottom by experts in the field and facilitated by superior athletes and coaches. The first graduates of that endeavor, members of the first and second battalions, bristled with vital energy, the kind of energy that eagerly sought expression, and found it in the form of song—

musical spontaneous combustion. It was a phenomenon, an unplanned, unforgettable event. We sang the Navy fight song, a cappella. Then again loudly, in unison. Then again in competition, one side of the gym against the other. The entertainment finished, we left the gym singing *Anchors Aweigh*. We sang it lustily all the way back to our three story dorms. Inside, up the stairs, our voices rang out, reverberating up and down stairwells. We kept singing after we had each reached our respective floor and then the competition started again, floor against floor, each trying to outdo the other in a form of a cappella warfare.

Our Battalion Officers showed signs of nervousness. They scurried around consulting each other. They looked baffled as if they didn't know what to do. This seemed like a scene that called for some kind of executive decision. But what do you say or do if it is not bedtime and the kids are just letting off steam, singing a patriotic song, the Navy's own? Are they singing too loudly? How can you sing *Anchors Aweigh* too loudly? But it did seem kind of rebellious. Nobody told us to sing. Is spontaneous celebration a breach of Navy Regs?

Not any more than letting butterflies loose in a submarine, I thought, back in the bar at the Officers Club. I smiled as I remembered what a surprising and unusual night that was. *Yet, those two battalions may have been ripe for just that sort of activity*, I thought. A lot of the guys had just finished college so they weren't really kids. Others, like myself, had finished the required two years of college. We were more sophisticated in the ways of the world such as manipulation of the masses through propaganda. We were more apt to think for ourselves and act on our feelings than were some of the younger ones, high school graduates, in later battalions. I had detected a touch of irony and even satire in that whole impromptu episode.

Back at the house, Roz had finished packing. We'd be leaving early in the morning. I had called my folks. They'd be anxiously awaiting our arrival. Our route was through the southern Cascade Range and the spectacular Willamette Pass, along upper lakes and down the clear flowing, white waters of the upper Willamette River.

As we drove along through that beautiful terrain, I sent glances at Roz to see if she was properly impressed. She looked pleased enough, but was verbally noncommittal. As I indicated points of interest, I expected some expressions of wonder and awe, but I heard only a few hmms and mff mff's from her. In all our years together she never admitted to the beauties of my home state. It was years before I figured out that she knew that if she should ever show even a hint of enthusiasm for Oregon, I'd start packing for the move.

In Portland, Mom and Dad greeted us. We visited with them for quite awhile, dined together in the evening, but I couldn't get Timberline Lodge off my mind. It was one of my favorite places in the world and I had been looking forward to returning there, especially to return with my bride. It had been five years since my last ski trip to the Mount Hood resort and to go back now would be another prize (besides Roz) for surviving the war.

In my college days I had skied on the nearby slopes and went to the Lodge often for lunch or hot chocolate. Now I could actually stay there with my wife in a *room, overnight*. I couldn't contain my excitement, and we started discussing the proposed trip. Roz had the ski togs we bought for her earlier in the winter at a shop on Balboa Island in Southern California. I knew mine were there in my old home along with my old skis.

I looked up the weather report for snow depths and conditions. A good blanket of snow still covered Mount Hood's higher slopes and good weather prevailed. It looked like we'd go next morning.

I sensed my mother did not share my enthusiasm for the prospects of our mountain excursion. Her face (which I could read like a map) reflected the harsh opinion she had of our decision to interrupt our visit with her and Dad. I told her it would only be a day and a night. But she insisted that our stay in Portland was far too brief to be cut short by such frivolity. She seemed to be directing her criticism to both Roz and me, as if Roz could call the whole thing off if she wanted to.

But the thought of the Timberline ambience had me completely entranced. Roz was totally innocent. I was the one so compelled to follow the dream, but it looked like my bride might take the heat for it. We did not relent.

The next morning we said brief goodbyes to Dad's warm smile and Mom's icy facade and set out for the wonderland of the Cascades. The last leg of the trip from Government Camp Village to the Lodge was a steep climb with numerous switchbacks that took us up quickly past snowy slopes into gathering clouds that turned into swirling snowflakes.

Roz, wide eyed and a little frightened, didn't realize how much I knew of our surroundings, how close we were to our destination. She thought we were lost in a blizzard. When we made our last turn into the lower parking lot, she wanted to know why we were stopping there. The falling snow hit her in the face as we got out. The Lodge was out of sight, leading her to think we were in the middle of nowhere. When I reached in the trunk to retrieve our gear, she thought we were about to walk off into the puzzling white limbo that surrounded us, when we heard a vehicle coming. It was the hotel jitney. The driver had come to pick us up. We climbed in with our gear and were on our way, on the last leg to Shangri La. Roz was still convinced we were headed into a swirling snowstorm abyss when it finally came into view: Timberline Lodge! It was a beautiful handcrafted product of the WPA in the 1930s, a building in which artisans expressed their talents in stonework, wrought iron, cabinetry, timbercraft, weaving, photography, and painting. You might call it Depression expression at its best and its crowning features were huge plate glass windows embracing views of the entire Cascade Range of Oregon.

I watched Roz as we entered the spacious stone and beam interior of the place. I could see that her face said "Wow" even if the words didn't. I knew she was impressed. As she moved close to the huge stone fireplace, a look of enchantment growing on her face, she warmed to Timberline Lodge.

Our stay was divine, our room, unique. A weaver in the genre of northwest native tribes had crafted the colorful window, bed and floor coverings; the hand-made furniture had carvings that reflected tribal decor. Because of the custom-made nature of the furnishings, no two rooms were alike.

We changed to our ski outfits. It was clearing outside and we had time for a go at the slopes before dinner. We rented a pair of skis for Roz and took on the snow bank below the lodge. Roz was a beginner. She spent most of the time on her back. As her instructor I had to tell her the first lesson after learning how to fall was how to get up. She caught on to the first part quickly, but didn't understand my lack of chivalry in the second. I tried to explain the importance of self-reliance, that you could find yourself in a position where there's no one around to help. Besides, it is hard to keep your equilibrium when the rescuee is off-balance while the rescuer tries to help her up. Balance is the secret and you can only attain it for yourself. Sounded like a good lesson beyond skiing.

Roz didn't get my reluctance to help. She thought me willfully cruel for making her lie on her back while she tried to place her skis downhill so she could get up easily. Nothing made sense to her. She finally figured it out and succeeded only to fall again at once. Oh well, time for dinner.

After dinner we happily relaxed over a brandy by the huge fireplace in the upstairs lounge. The downstairs lounge was the one I visited in my college days. It was plenty nice but this was downright luxuriant. There certainly were advantages to being a bona fide guest there.

The next morning we tried some more skiing with a little more success than the day before. It would take more lessons. We checked out at noon and headed for Portland with reluctance and no small apprehension. It felt like storms ahead and I don't mean the weather. The patterns of Mom's displeasure were easily remembered and I knew she wanted the impression to be lasting. With that in mind I suggested an early departure for Palo Alto. It was urgent that we find out if Stanford wanted me or not. Our future rested on that knowledge.

Back in Portland, we told Mom and Dad of our plan to leave early the next day, of the need to know where we stood in the admissions process at Stanford. We apologized for cutting our visit short and hoped they understood. Oh, yes Mother understood the whole thing only too clearly. I sensed a kind of psychic battle going on. Roz and Mom were both visibly upset and yet neither had said much; it was all rumbling under the surface. The new little wife had somehow taken control and would undoubtedly pay dearly in the future.

A couple of days later we were back in Lafayette. I made an appointment with the Admissions office at Stanford. They told me the Registrar would interview me. He would break the news to me. I dressed up this time in my wedding costume; the gabardine khakis with silver bars, wings of gold, and a band of ribbons including the Purple Heart. This time I added my Sam Browne belt, a leather belt with a shoulder strap through the right epaulette, that exuded an air of authority, suggesting some kind of special achievement. Of course it was merely a piece of optional equipment. But would the Dean know that? I didn't know. Every little bit would help.

At the administration building I met with Dean Winbigler, the Registrar. He was a friendly, professorial, middle-aged man who looked me over. I hoped he noticed my gold wings and the Purple Heart ribbon. He told me he couldn't admit me as a regular matriculated student because of grades and credits from my first two years of college at the University of Oregon where I had majored in Physical Education. Most of my higher grades there had been in P.E. lab courses and they didn't count at Stanford because they had no such major.

I slumped in disappointment, clenched my teeth. It was a real blow to my self-esteem. The dean, acknowledging the effect of the news, studied me for a moment.

"You obviously have achieved much since your two years of college and for that reason I think I can offer you a proposal," he said. "If we enroll you in the summer quarter you will have an opportunity to prove yourself. If you take a full complement of courses and get outstanding grades we will reconsider at the end of summer and enroll you as a fully matriculated student. But, I mean your grades must be outstanding. How about it?"

"Wonderful, sir! Thank you so much." I was thrilled and exhilarated at the encouraging prospects.

I could hardly wait to get back to Lafayette with the good news. This seemed to be another example of the help from a higher place I had been so lucky to receive ever since that big bang on February 10, 1944.

While I was at the office I inquired about housing, again. I had heard hints of some married student quarters someplace off campus. They called it Stanford Village: a former military hospital (Dibble Army) in nearby Menlo Park had been converted into a series of one-bedroom apartments for married students. There was a waiting list. I decided it would be a good idea to sign up at once. There probably wouldn't be a vacancy until next spring's graduation, but it would be worth the wait. We still had our temporary room in Palo Alto, which would be ours on June first.

CHAPTER 17

BACK TO THE DRAWING BOARD

When we moved into the house on Fulton Street, our first move of many more to come, we finally got to meet our new housemates. The young Stanford couple was delightful and engaging, a pair who could become lifelong friends. Don, a graduate theology student, was studying the Chinese language in preparation for a career as a missionary in China. Helen had a position in Stanford Memorial Church on campus, as an apprentice organist and could play the huge pipe organ that filled a vast portion of the upper gallery of the church.

When we were alone they gave us the lowdown on the living situation there in our new home. The landlady had implied that our stay here was in some way linked with her personal needs. Her young son was a victim of something like cerebral palsy and because she worked, she had difficulty making the other parts of her life come together. Helen had found herself doing unexpected chores and being engaged to perform them at inopportune times, making it hard to get her own schedule underway.

Although there had been no formal agreement linking rent,

payment, and help, Roz soon found herself tangled in this web of activity along with Helen. The two of them were confounded and soon energized into a fresh search for simpler living quarters.

The climax came one day when the woman asked Roz to help with preparing a meal for a dinner party. Roz sort of got caught up in it because of her own interest in cooking and would be working solo because Helen had a schedule conflict at the time. She even confided to the lady that she could get groceries and meats at the commissary on nearby Naval Air Station at Moffett Field for a discount, and they both knew about the general meat shortage. Roz spent the whole morning shopping, the afternoon in preparation of the meal, and when she thought the end had come at last, she was asked to help serve and to wear one of those cute little French aprons. The fun part of the fiasco had long faded by then, disappearing completely when she found herself clearing the table, then building to anger as she washed the woman right out of her life along with the dishes.

No doubt we'd be out of there on July First. Don and Helen had already found another place. My classes had started in the middle of June so Roz went out on her own every day, intent on finding a place before the end of the month. A vacancy showed up—just in time—during our last week on Fulton in the form of a whole house, furnished, that we could housesit for the summer, all by ourselves. That house was on Alma street, the A street closer to the campus than the F street. *Could it be a good omen for my classes?*

I was enrolled in courses where I really felt comfortable and confident: Art classes called, *Beginning Drawing, Decorative Design, History of Motion Picture Making,* math and engineering classes, *Foundry Practice,* (my father's profession) and a gym class in *Handball* for exercise. I felt I could do well in the foundry class since I already knew some of the nomenclature and the lingo. By the time we moved I was two weeks into school and feeling confident.

A week or so after our move into the new place I arrived home after a late afternoon class to find Roz waiting for me at the gate with a smile on her face, one hand behind her back.

Her smile broadened as she gave me a cute little salute, "Hi, Captain."

"What's with the Captain?" I asked.

As she gave me a quick kiss she brought an envelope out from behind her back and said, "The answer's here."

It was one of those official looking brown envelopes from the Commandant of the Marine Corps, addressed to Captain J.G. Morris. I opened it quickly to read the official word "It is!" A totally unexpected promotion to Captain "... for satisfactory service in performance of your duties as a First Lieutenant."

"I swear, someone at headquarters likes me." *Or was it another act of divine benevolence anointing me again?*

"Maybe you just earned it. You had it coming. Let's go celebrate!" she said. "We'll go to Rickey's."

"Hey, you're beginning to sound like one of the natives around here." (She meant Rickey's Studio Club over on El Camino Real, a legendary landmark among Stanford people.)

Two weeks later, another one of those official brown envelopes arrived from headquarters in Washington with not-so-rosy news. First it told me that the President would sign my orders on August First. But to be official, they would have to retire me as a Second Lieutenant, that being the rank I held when wounded in combat. The letter officially demoted me. Three weeks ago they called me First Lieutenant; two weeks ago, Captain. Now I was a lowly Second Lieutenant again and I was destined to be one for the rest of my life. Ah, well. At least I was lucky to have a life.

I seemed to be making good progress in the grade department. I had just gotten a paper back with an A on it: a shooting script I made for the motion picture class. My drawings were working out really well. I was having no problems at all with the foundry class, getting "A's" in Algebra. I was confident we would be there at

Stanford University until I graduated.

Roz was motivated to keep up her quest for more permanent quarters. Competition for housing would increase when fall quarter started, so she hit the pavement again, this time along the lanes of Atherton, a pricey neighborhood of huge estates just west of Menlo Park where our new friends Helen and Don found their place. There just might be more of the same: empty living spaces just waiting for someone to ask. Roz had the chutzpah to do just that.

She searched for empty servant's quarters, gardener's cottages, chauffeur's garage apartments. She knew what to ask for and she did. At the large estate of the Pullmans of sleeping-car-fame, there was no answer to her knock at the main entrance so she walked around to the rear and spied a separate building that appeared to be a place suitable for the help. Sounds of human activity inside encouraged her to knock. A friendly looking woman in a housedress answered and Roz told her what she was looking for. The woman invited Roz in and introduced herself and her husband as the Pullmans. They were comfortably settled in their own servants' quarters. The mansion was too big for their purposes, too much for them to maintain and they seemed to enjoy the hominess of the smaller quarters they now occupied. They believed they would eventually move back into the big house when help was more available. They praised my wife for her determined effort to find a place for us and wished her well.

On another foray into the wilds of Atherton she stumbled upon the estate of an ancient baseball player named Ty Cobb. When he answered the door, completely bald and dressed in a brown bathrobe, she mistook him for a monk. When she asked for the owner, he politely informed her he was the owner, whereupon she told him her mission.

He came out and proceeded toward a side courtyard. Roz followed through a gate and he pointed out a cottage alongside a swimming pool. "What a shame you didn't get here a couple of days ago," he said. "I just rented this to a pair of college kids.

They're moving in today."

A happy looking twosome popped their heads out the open door and Ty introduced them. They told Roz she was on the right track and encouraged her to keep looking. "There is a place out there for you. You only have to ask." She thanked them for the encouragement and congratulated them on finding such a nice place.

Her search would bear fruit soon. She felt it in her bones as she walked down another long drive with visions of an empty apartment at the end. As she rang and waited she looked around for signs of servants quarters. She saw windows in the space above the huge four-car garage.

"Hmm, that looks promising up there," she thought.

A middle-aged woman answered the door with a smile. When Roz asked if the garage apartment was available as a rental, she showed a sudden interest.

"Well, I hadn't thought about it but it is empty. We don't have a chauffeur any more. Uh, could you give me some time to think about it? Call me tomorrow. I'll give you my number."

Roz took a card from a hand that appeared to have a slight tremor, thanked the lady for considering the possibility, and hurried back down the long drive with hope in her heart. "Maybe my persistence is about to pay off," she thought.

She looked at the card. It was actually not the woman's card but her husband's. He was the owner or president of a large transcontinental trucking company. The woman had written her home number on it.

In the evening, with great excitement, Roz told of her encounters with the Pullmans and Ty Cobb, but proceeded a little tentatively about the prospect of the estate with the empty garage apartment. Something about the woman's sudden interest and friendliness had an undercurrent of oddness about it. I congratulated her for her gutsy ventures into the land of the privileged and tried to reassure her that the woman had only awakened to thoughts of good deeds.

The next day Roz made the call. The woman's interest had risen to another level. She wanted us both to come and look at the apartment, for she was sure she could offer it to us. When we reached our destination and proceeded up the long drive I could see the wide apron in front of the large garage off to the left. This was a convenient location. No need to go near the main house, and large shade trees almost shielded the apartment completely from view. It could be an ideal setup.

Our potential benefactor greeted us like long lost friends, invited us in, offering drinks. I noticed she already had hers poured and partially consumed. She could hardly wait to break the news that we could have the apartment and we were such wonderful kids it would be free. She didn't want to deal with finances. This sounded a little strange. She hardly knew us. She was ready for a new life with two strangers. Did she need friends? Were we it? She hid her drunkenness quite well; it was her irrationality that gave her away. We went along with the formalities but we knew it was a no-go deal. We looked at the apartment, visualizing how nice it would be to come home to, swallowed hard, and told her we'd call—which we never did.

Finally, the long awaited retirement orders arrived. The papers did not have the President's signature but they were authenticated by the Commandant of the Marine Corps who announced further good news in an accompanying letter from Secretary of the Navy Forrestal. He promoted me back to Captain on the retired list! I wouldn't be a Second Lieutenant the rest of my life!

Further good news came with the retirement orders. I had forgotten about a recruitment bonus used by the Navy to increase the number of Naval Aviation Cadet candidates who enlisted early in the war. They had offered five hundred dollars a year to us for each year we served on active duty. A check for two thousand dollars came along with a retirement that allotted me three-fourths of my base pay each month for the rest of my life, "space available" air travel on military planes, base privileges in commissaries and exchanges, plus medical care at military or civilian facilities for me and my family.

Because of my injuries I was guaranteed complete rehabilitation at a college, university or training institution of my choice. I had made my choice: Leland Stanford Junior University, and it would choose me when summer quarter ended.

Rosalind and I are eternally grateful to Uncle Sam, to Stanford and to a certain mystical Presence for some of the best years of our lives.

EPILOGUE

EPILOGUE

ASHES

I had come to a difficult time of my life to think about and record. This was the day of Rosalind's memorial service, November 16, 1990. We—our grown children and I—had invited our friends, old and new, along with relatives, to honor her life.

She had succumbed to a rapid cancer growth only six months after its discovery in April. Now that I had my kids here along with old friends, the pain of the loss was softened and postponed. We had been busy with the "arrangements."

While singers were recalling some poignant heartfelt tunes from our past, some of our acquaintances had questions on their faces: "We didn't even know she was ill." "She seemed always healthy and vibrant." It was hard for some to celebrate her life while they were still shocked at her death.

A woman minister, Joanne, from the Church of Religious Science, spoke about the many roles Roz played in the theater of life: as wife and mother, in dental assisting, as a bridal consultant, and volunteer in fund raising as well as public relations work that grew into a full-blown profession with a university certificate and membership in a women's professional organization. The minister related how Roz had blossomed after the children—Jane, Wendy, Ginger and Tom—had grown and flown, how being a mother of four was only the beginning.

Spontaneous eulogies came from important people in her life. Patricia had flown from Stanford to tell of her own personal relationship with Roz in her work as a fundraiser.

Her nephew Richard told how fragrances from her kitchen often meant delicious soup with savory flavors found nowhere else on earth.

Nancy told about the time Roz had been flown to the campus for a special event because of her excellent talent in fund raising. The Stanford Office of Development brought people like her from all over the nation for this occasion. At the evening dinner, guests were seated with name cards at their places. When Roz found hers next to the President of the University, her mouth flew open. "This cannot be," she exclaimed, "They got the wrong Roz Morris!"

Her friends heard her and reassured her that she was the right one, that she was meant to sit there. (There happened to be another woman with that name who did volunteer work at the Children's Hospital and Roz had assumed the place was meant for her.)

Women in Communication members, fellow workers, friends, old and new schoolmates, told anecdotes about other facets of her life. I paid homage with the final remarks. I emphasized my feelings about her loving service as a performer in the theater of life because of the many roles she had played. I suggested we take a minute to give her a silent standing ovation in our hearts.

I told them that a final resting place had not yet been decided for her, a person who never seemed to rest; that in the meantime they could think of her whenever they found themselves in a remarkable part of that great big beautiful world out there, and I gestured toward a large picture window behind me looking out on a garden.

Inspired by poignant lines of Tom Joad in "Grapes of Wrath," I said, "Wherever you go. Whenever you find yourself surrounded by beauty, think of her. In Paris, strolling down the Champs Elysee, through the Tuilleries, into a great museum like the Louvre or Musee D'Orsay, think of her, she'll be there; in any place of great natural beauty like the Pacific Coast or the California mountains. Think of her wherever good people get together to do good things. She'll be there."

We really had not decided what to do with Rosalind's ashes. They were sealed in a ceramic urn at the funeral home awaiting our instructions. The week after the memorial service we gathered at Wendy and John's for Thanksgiving in Albany near Berkeley. On a hike we found ourselves clustered on a huge boulder called Indian Rock in the Berkeley hills. We remembered we used to call our family meetings "powwows." What better place for a powwow than right there on that big boulder? We could decide what to do with Rosalind's ashes. We wanted to do something significant, not tuck them away in a crypt in a cemetery. Something about her youthful beauty, especially pronounced the night she died, made us want to keep her memory alive a long, long time; seemed to call for something special for this woman who didn't want to grow old and never did. We wanted to do something lively.

Someone suggested we do what I had hinted at in the eulogy: take her ashes to beautiful places anywhere in the world. We could divide up her ashes into small containers, one for each of us, ready for scattering discreetly wherever it seemed appropriate. Reacting to the emotional impulse of the moment, we all agreed enthusiastically with the concept. We would find some fancy little bags with drawstrings and I would pick up the urn, fill each bag, and give one to each.

We talked about places we could do this, trips we might take. Fortunately Tom spent a lot of time in France where he served as Producer of Fantasyland at the new Paris Disneyland then under construction. He could return to places that Roz and I visited just a year before: Monet's Garden in Giverny, Chartres Cathedral, her

Chateaux in Loire Valley, Mont Saint Michel, Pont Neuf Parc on Ile de le Cite, in the hub of the City of Light, Paris.

Closer to home, Bixby Creek bridge on the Big Sur coast, Point Lobos, Carmel Bay, were all places she loved; Jasper Ridge on the Stanford Campus, San Francisco, the Golden Gate. The world is huge. We could pick and choose. And we did.

The week after Thanksgiving, less than a month after she left us, I found myself home alone. I felt lonely and sad. Half of my work was gone with my love. Empty places abound all around, her place in bed, sounds, and fragrances. Well, not completely—I sometimes opened her closet, stepped in, and gathered an armful of her garments to hug and smell. Much of her was still there but not enough. The empty places stood out, especially the one in the center of my chest, the one that announced itself with a dull and growing pain. Heartache had only been a word to me. Now I knew its reality.

I looked for her in places outside of our Newport Beach condo, in the patio, an outdoor private place, then farther out into "our" world where we took walks, then farther into the nearby nature preserve in the back bay. I looked for her face in the clouds just as I did those many years ago in the South Pacific, on that St. Valentine's Day when she heartened me with words of hope. I needed to tell her I loved her. Out on the empty playing fields of the high school, a brilliant sunrise helped me think she called back.

I found myself shouting (legato, under my breath), "I LOVE YOU!!"

I couldn't shout it enough. I had to make up for the times I had taken her for granted. Too many times. Way too many.

She left too soon, before I was ready, and she took away my job as a caregiver. My focus on her survival had lost its purpose as I wandered aimlessly to places we used to go, watching other couples of all ages, under my breath, urgently, telling him, "Hold hand," or "Smile at her," or "Be her lover, you may lose her!"

If they parted, went their separate ways without a kiss or a hand squeeze, as I followed along the promenade, I shouted (again legato, under my breath), "YOU FORGOT SOMETHING, DUMMY!"

I had suddenly taken on a role as silent director of partner behavior. I didn't want anybody to be taken for granted anymore, a sort of new twist on old scenes from "Our Town," the great classic play by Thornton Wilder in which Emily, the heroine, dies young but gets to relive a day of her choice, a brief second chance. I sensed a parallel between that tearjerker and my present feelings. I remember the first time I saw the play, so full of feelings it brought tears to my eyes, a lump to my throat. It took a few weeks for it to dawn on me that whether I liked it or not, I had reached a crossroad in my life.

I was back on the road and in the classroom after the holidays, partly back to routine in my role as "freeway professor" at various community colleges and design schools. After each class, I felt my automatic homing beeper turn on. If it were electronic, I could have just thrown it away. The only way I could deal with it was to have a dialogue.

"Why are you beeping?" I asked.

"Time to go home."

"What for?"

"You know, the usual."

"There is no usual anymore."

"Oh, yeah... so what are we going to do?"

"I dunno, but if you'll stop beeping, I'll ease up on the gas."

The last five years before cancer struck her down, Roz and I had gotten into an easy routine in which our roles were slightly reversed. She had become a part-time breadwinner, while I had taken on home maintenance chores, cleaning, laundering, cooking. Her late shift at the Library brought her home at seven in time for a well-prepared meal. It was not easy cooking for a woman with the reputation of a fine and creative cook. Those earlier years when I had served as her helper gave me the confidence to take on this auspicious position. Occasionally I would do a little improvising, a sprig of cilantro along with the cumin to change the zip to a zap, something to give the dish an extra edge. These slight deviations would often bring admonitions from the master. Any changes in proven recipes called for careful thought.

It was hard to get used to the fact that no one expected me home. I could stop anywhere, stay as long as I liked. I continued my activity in my alumni group, writing the quarterly newsletter, part time teaching, helping with an installation in a local museum. At this crossroad in my life I had plenty to do as I sought direction. A class in Tai Chi Chaun helped me find direction. Later in the winter a brochure from the Stanford Alumni Association Travel-Study group arrived announcing a trip to the South Pacific that would start in the harbor at Rabaul. It would follow the route of my return from Rabaul during the war, stopping at the island where the rescue plane had taken me to the seaplane tender, USS *Coos Bay,* in the Treasuries. It would make stops at Guadalcanal, Tulagi and Espiritu Santo, and other historical sites along the way. I had to take this trip!

Besides seeing all the old places again, which would be a thrill, the thought of again seeing a part of my escape route and a sighting of the crash site had me absolutely spellbound.

I had daydreamed many times about going back there, finding the crash site, and bringing home a fragment of my plane. This trip might afford the opportunity, might make the daydream come true. Even more important, I could take Rosalind's ashes to one of the beautiful places in the South Pacific. I sent in my deposit at once to secure my reservation. The dream marched on.

On an old map left from combat days, I had already drawn what I thought was my escape route and plotted the possible crash site. The clues were on the map and in my memory of the hike.

★ ~~~ ESTIMATED CRASH SITE AND PATH
✱ .... ACTUAL CRASH SITE AND PATH OUT

Landmarks. Now I was actually going there. I would see if I was right. That well-trod path I crossed on my third day in the jungle compared well with the one on the map, coursing NNE, connecting villages just southwest of Vunakanau, our target that day. I saw no other path with that configuration, with that juxtaposition to the target field, but I wouldn't know until September.

Meantime, the classes I taught in Architectural Drafting and Interior Illustration, and helping install an architectural exhibit, would occupy me before the trip.

When the time came to make reservations for the flight, I decided to arrange for arrival in Rabaul two days before our ship's embarkation date so I would be sure to have time to make a search. As to Rosalind's ashes: I felt the time and place would present itself, as the adventure unfolded.

My Quantas flight took me to Brisbane, Australia from Los Angeles via Honolulu. After an overnight stay in Brisbane I boarded an Air Niugini flight to Port Moresby, which took us high over a long stretch of tiny atolls and submerged shoals, revealing the vastness of the Great Barrier Reef along Australia's Northeast Coast, west of the Coral Sea.

At Port Moresby, Papua New Guinea, I felt as uneasy going through customs as I had at Brisbane. I was concerned about that small linen bag with drawstrings. *Would they question its contents if they found it? I could tell them it was a supplement I take, calcium. What if they asked how I took it? Would they want a demonstration?* I managed to get through both times without detection, but this would be a concern throughout the trip as we passed through small island nations. Nevertheless, I had my eyes open for the beautiful place the ashes would finally come to rest.

We transferred to another Air Niugini flight, which would take me to my destination: Rabaul. I anticipated the arrival over Gazelle Peninsula with contained excitement, handycam at the ready, prepared to get aerial shots of the rivers, volcanoes, bays, and harbor. Normally, the prevailing winds would have us approach

Epilogue: Ashes

the airport (Lakunai airfield during Japanese occupation) from Blanche Bay toward Rabaul town and Simpson Harbor. I expected to get good shots of Cape Gazelle, Duke of York Island, Warangoi Bay, and other wartime landmarks.

As we approached the peninsula, the cloud cover below increased. When we got there, the landmarks were nowhere to be seen. We approached the field from the other side—the Kerevat side— and let down through the clouds, banking around a volcanic peak and slipping in between a ridge and the overcast, over the town of Rabaul from the north and into the airfield heading southeast. It was more like a sneak attack than the grand entrance I had envisioned.

I only had time to check into the Travelodge, no time to do more than dine, then hit the sack and dream about what awaited next day.

Early in the morning after a refreshing stroll along the waterfront, I set out to find maps and information about car rentals, guides and all. I wanted to find a detailed map of the extended area south of Rabaul and to see how close I could get to the Warangoi River mouth, as well as the area just south of the old Vunakanau air strip we were bombing the day I was shot down. I presumed I had come down five to seven miles south of the field inside a diamond shape formed by connecting paths that could have become roads. I surmised that I could drive out there, park, and scout through the terrain to see if it looked familiar.

I found a tourist bureau in town where I expected to get the map that would open those vistas for me. But alas, their map matched the one I found in my hotel room. They said if I wanted a really detailed contour map, I could get one at the Land Office, which, as it turned out, was only a block from my hotel.

At the Land Office they had posted bad news: "We are closed 'til Thursday," the day we were to embark on the ship for the grand tour of the Solomons and beyond.

A short time later I popped into a travel agency along Mango Avenue, the main drag, in hopes of finding a better map than the one I had. Through the course of a conversation with a young woman there, I learned she could get a driver/guide for me and the best place to rent a car happened to be right there by my hotel. She told me it would take some time to locate the driver, so I used the interval to walk through town, a stalwart community that had been rebuilt more than once. Rabaul had survived bombings and volcanic eruptions. It was incredibly resilient and alive with activity. I sensed a great optimism from its people, especially at the large open-air market. Its huge array of colorful native fruits and vegetables from surrounding farms was displayed in booths under dark canvas. This square block market, ringed with late model trucks and vans, reflected a surprising level of prosperity.

I used my handycam to record places I visited as I waited: the War Museum at Yamomoto's Bunker, the new Rabaul High School.

When I finally returned to the travel agency I found that they had contacted the driver, Stella, a young woman who knew her way around the area. At the car-rental office, I found that they knew Stella and were already contacting her as part of the deal. Meanwhile the owner of the agency, Phillip White, heard me talking about my mission and entered into the conversation. He heard me say something about a search for a downed aircraft. It jogged his memory about a collection of war relics on the island. There were people in the area who had taken part as volunteers in the gathering, preservation, and display of these pieces. He handed me a card with the name and phone number of Alistair Norrie, a volunteer at the Kokopo Museum who might have some documentation on aircraft wreckage.

With travel arrangements completed Phillip invited me for beer in the lounge where he introduced me to other locals who had tales of great interest. We swapped war stories for an hour or so. Then off to bed I went.

The next morning Stella arrived, and Phillip had upgraded my car to an all-terrain-vehicle in anticipation of narrow, bumpy roads, no roads, or... what? I called Alistair Norrie's number at 0745 and he answered. I told him about my mission. He sounded genuinely interested.

He asked, "What type aircraft was it?"

I answered, "F4U-1, Corsair."

"Do you happen to know the bureau number?"

I happened to know the number very well. It almost matched my Marine Corps number.

"I know it very well. It's 02566," I replied.

Stunned silence on the other end. Then, 'We have it! We have your plane in the museum at Kokopo! Can you meet me there at 0830?"

I felt the hair on the back of my neck rise, a chill went down my spine. *My God! Is this really happening?*

I was so excited, I almost shouted my answer, "Yes, I'll be there!"

I checked out of the hotel and loaded my gear into the ATV because I would be going aboard the cruise ship at 4:30 p.m. We were on the road to Kokopo by eight. The drive was less than a half-hour.

I was filled with wonder, surprise, and anticipation. I had expected my mission would take more effort than a mere phone call. At 0800 I was already on my way to visit the ghostly remains of a long-ago, traumatic incident.

I suspected my driver wanted to prove her prowess as a tour-guide as she pointed out places of historical interest along the way while my only point of historical interest right then lay in a museum twenty miles away. She even stopped near the western shore of the harbor to bring to my attention another of the small volcanoes that erupted out of the giant caldera that formed Blanche Bay. Vulcan had first risen as a cone shaped island in 1937, then grew into a peninsula as it joined the rim of the bay. I must admit it was a spectacular view of the whole area. I took a few quick shots with my video camera and urged Stella toward our real destination in Kokopo without further delay.

Finally, we reached it: the Kokopo War Museum, a low wooden barracks-like structure and a metal shop/warehouse all behind a chain-link fence through which various war machines lay exposed, open to public view and the elements just as they were in action many years ago. At the gate Alistair, a middle-aged man, waited with a younger man whom he introduced as David Lindley, a geologist, and a young native islander named Frank.

Alistair explained David's presence. He and his crew found the remains of my plane out there in the wilderness bush while on a geological mission in search of gold and other valuable minerals.

We didn't talk long. Something more impelling lay ahead as we approached banana trees on a slope, and there it was! The silvery remains of "my" old Corsair. It lay thoughtfully arranged with the plants growing up through it. Alistair explained its location outside. They had decided to display the plane as they found it in the jungle, crumpled hunks of metal, as it had landed, upside down, wheels in the retracted position. There she lay, as I last saw her, 48 years ago, when she was aflame and crackling from her ignited machine gun shells.

As David and Frank lifted a section of the port wing, Alistair explained, "The main tank just kept burning in and around the cockpit area."

Under the port wing there was inscribed, in pencil, BUREAU #02566 along with the date of the finding and the name of the official, Brian Bennett, who identified the wreckage by finding the three places where enough of the numbers survived to put it all together. Brian was the authority who later applied for permission to put the relic on display.

My escorts seemed to sense my emotion and quietly drifted away as I walked slowly around the wreckage, camera in hand, lump in throat, head shaking in disbelief. The emotion I felt was indescribable.

 Later, inside, they showed me pictures of the crash site, related how they found the plane on a slope, low in a valley near a stream, unseen and untouched for 42 years. All six machine guns and the ammo were still there. This is a good indication of how remote the crash site was, truly a wilderness area, a good twenty

kilometers below the spot I had originally planned to search.

Many times in the last 47 years I had said that some day I would return to Rabaul and find my plane. I was right but I didn't know others would do all the hard work for me. After finding the plane devoid of human remains, they had decided to remove it and display it as an example of how a crashed aircraft would appear out in the bush and to honor all those pilots and crew members who didn't make it out alive. The required approval of governmental officialdom took almost five years. Only 9 months before, they had hauled it out in a net by helicopter, minus the engine and propeller, to the nearest road, then by truck to its present location. My timing had been perfect. The engine and prop are still at the crash site, too heavy for the little chopper.

Curious about some other details of the crash and its aftermath, I asked David about the wide, well beaten path I had crossed before reaching the river. He told me about a wide trail the Japanese had made between two radio outposts on two knobs below the Warangoi River just east of, and overlooking, the crash site. That appeared to be about as close as I came to the enemy before I reached the plantation on the coast.

I learned a lot that day. The little river that took me to the coast has two names: Sigule and Sicute. The natives call it "Sicut." David had a detailed land map and it showed this little river joining the larger Warangoi just before reaching the coast. That identified it for me because on the day I escaped from the island, I waited for darkness at the confluence of those two streams.

On the map I could see the exact crash site, the two knolls that had held the enemy radio stations and the little creek where I had washed my wounds. Even this little stream had a name, Magut. I got far more information than I had anticipated, and then David offered to take me in his ATV down to the river mouth, the beach at Warangoi Bay. Of course I jumped at the opportunity.

I asked Stella to pick me up at the museum at mid-day. We would finish our tour at that time. Until then that amazing experience continued to expand as we made our way toward

Epilogue: Ashes

Warangoi Bay.

It had all happened so fast. At 0745 I learned that "my plane" had already been found, its old bones resting comfortably in a museum; by 0830 I walked around it in disbelief and at 0900 rode a narrow, bumpy road toward another emotional encounter.

At one point as we descended a grade, David stopped and pointed out some landmarks in the middle distance, off toward the far away Baining Range. He indicated a knob that showed on the map at an altitude of 568m. Over a ridge and just to the left of this lay a valley, the crash site, over twenty kilometers away, accessible only by helicopter or a three day trek through rough terrain and dense jungle.

A short time later we could see the same landmarks from a bridge over the Warangoi. We had now reached a point less than two kilometers away from the confluence of the two rivers. As we drove the road between them, we approached a pivotal point in my long-ago trek to the sea, a literal turning point in the middle of the sixth day when I had rounded a bend and saw the plantation dead ahead. Another poignant moment was imminent.

As we descended another grade, David informed me we were on our way down to the Sigule but I couldn't see a thing through the thick foliage along the road. No long view of the stream, just a sudden turn around a clump of tall reeds and our front wheels suddenly came to rest in the water. We had arrived. No doubt about it. On the right was the bend in the river that I came around that day and suddenly saw the plantation now off to my left, the stream plunging straight toward it, then veering abruptly left and out of sight around another bend. Everything was the same, except the streambed had changed course slightly to the left as it approached the bank below the grove. It seemed as if no time had passed since I hid my raft in the reeds and climbed up that bank to find coconuts for the anticipated sea journey ahead.

I took off my shoes and waded around in the little river as a wave of gratitude swept over me for the comfort it had afforded me as it aided my journey to the safety of the sea. I found myself seeking a ritual of thanks and appreciation: a tribute to a tributary. This private ritual went on inside my head and in my heart as we drove on down a narrowing lane toward the beach where soft waves of St Georges Channel met the current of the Warrangoi. This was the place when, on that lucky night many years ago, wind and current took over and carried me to my point of rescue two days and thirty miles later.

My dream to find my plane and visit the little river came true on that lucky day. My watch said 1030 and all this had happened since 0745, less than three hours... incredible! I thanked David for taking his time to "gift me" with that memorable return to a traumatic and dramatic piece of my past.

On our return to Kokopo I thanked my newfound friends for making the dream come true and surprised them with a copy of a story I had written about "the plane that almost had my number." They had a surprise for me, a gift that completed the dream. They handed over a piece of my plane to take back with me, a piece they had kept in the warehouse from inside the cockpit: a switch panel—a piece I had touched many years before now touched me in a profound way. A gift to top all gifts!

I let Stella continue her routine to another landmark: the old Vunakanau airstrip, unused but still visible, now overgrown with plants. Then we went south to a viewpoint high on the hills where the knobs that had held two Japanese radio transmitters surrounding the crash site south of the Warangoi River course were again visible. The whole panorama, the landscape of my adventure, lay spread out before me.

At the museum I had learned about another plane that had crashed in the area south and east of that river. The sole survivor of an Army Air Corps bomber had traveled west, crossed the river and found refuge in a village. Friendly natives hid him from the Japanese until word got out through the grapevine; they found him and took him to a prison camp where he stayed until the war's

TRIBUTE TO A TRIBUTARY

The labor lasted four long days—13 through 16 of February, 1944—on your little river. A gift from God's Angels, to use as a rebirth canal. They reserved it for you, swept it clear of danger, left some mist for mystery; a stray hog strayed in; your error to think: *it's food and the mist is smoke.* speckled trout and crawdads for reassurance. The water, cool, pristine, sparkled and gurgled yesses to questions of safety. And what a river—a stream named Sigule whose twin, called Canyon Creek near Mt. Tum Tum, thousands of miles away taught his boy while Dad cast flies for rainbows as his sprout dashed about up and down banks, across boulders, monkeying around on low arcing maples, gazing like Narcissus into reflecting basins; pausing, calmed by quiet, lazy eddies over deep blue pools. A stream resembling the recent rebirth passage.

Here, you grew, sad, stressed and gloomy—wondering why it took so long. Feared you'd never get out of there. An Angel noticed it was St. Valentine's Day, that you could use some heartening, brought in a loved one from far away who sent thoughts of devotion, inspiration, love. Your eyes widened at the mystery of the brief visitation. Instantly a warm infusion flowed through your being, lifted you, gently urged you on. At last, with the ashes of your Phoenix, six days back, your rebirth was at hand. You simply waited 'til darkness fell, slipped into your tiny craft, into the swollen little Sigule which bore you to an ample Warrangoi, then around its last bend to feel the rise and ebb of the Pacific. Rebirth was at hand and you rejoiced as you were borne into a nourishing, dark abyss; the swaying, sheltering arms of Saint Georges Channel at night.

end. That could have been my fate if I had turned west.

At 1600, as Stella and I approached the town of Rabaul we saw that my ship lay moored at one of the docks; M.V. *World Discoverer* awaited passenger embarkation. She is small, accommodating only 152 travelers, which makes her extremely maneuverable in shallow waters. Passengers can go ashore in many places on small Zodiac boats that drive right up on beaches to native villages and to reefs for snorkeling.

I took my traveling gear aboard and met tour leaders and some fellow passengers. Back in the car I told Stella that I wanted to get one of those detailed maps at the land office, but they were closed for painting.

She said, "No problem. I know them. I'll get your maps."

At the Land Office she walked in and returned shortly with the good news, "They said yes." What a lucky find was this young woman.

Maps in hand, I returned to the Blue Star Hire-a-Car office with news of my morning discovery. I thanked Philip White for his part in my finding the remains of my plane. If he hadn't been alert and overheard my queries the day before, I might never have found it.

Back on board, while fellow passengers went ashore to explore the amenities and the sights of Rabaul, I felt satiated beyond fulfillment and content to stay out on the deck of *World Discoverer* to drink in the wonder of the day's fantastic events. My gaze followed the skyline over this giant caldera that forms the harbor and Blanche Bay, first the small volcanic cones rimming the town to the north and east, then focusing especially on the hills to the south that overlooked the vast wilderness jungles of Gazelle Peninsula. Feelings of gratitude encompassed me as I thought of the great good fortune I had experienced from the moment I tumbled out of that crippled war bird so many years before. I had made new discoveries this day. I had learned my plane hit the ground upside down only seconds after I fell out of it, that it made a hole in the jungle canopy on the way down. Was this the same hole I fell through? I dunno. The wonder of it all makes me think,

Epilogue: Ashes

Angels.

Speaking of Angels... thoughts of Roz floated in. Images of her appeared in the clouds again just as on that long ago Valentine's Day. *She's with me again...in spirit.*

Then I remembered the ashes. *Some beautiful place out here will receive them... I could place them tomorrow in St. Georges Channel on the spot my rescue took place. Yes... that's it. We should be there at sunset tomorrow... perfect!*

On reading the itinerary for the first evening aboard, I saw that it called for a visit to a native village where we would witness a ceremonial fire dance. It didn't state the destination, but I soon learned we would board buses and head off toward Vunakanau, to a village southwest of the old strip.

"Say, that's down in that small network of roads and villages where I had planned to start my search this morning, before the phone call."

After this busy eventful day a native fire dance would be only commonplace. Besides I felt bushed from excessive adrenaline flow and I dozed as our bus meandered southward along narrow roads.

The ceremony unfolded with brilliant masks on zealous actors telling stories of love, courage, disappointment, jealousy, and unbounded happiness, universal emotions found around the world. Each scene called for a jaunt by players in grass skirts through a crackling bonfire whose flying sparks kept the audience awake and alert.

Afterwards, sleepy travelers boarded buses for the return trip to the ship. With their engines running, we sat and waited... and waited... wondering, "Why aren't we moving?" Then word got out, "We're waiting for our police escort!"

"Police Escort!?"

So, the warnings I had heard earlier in the day weren't idle rumors.

They had spoken of "Rascals" who haunt roads in the bush,

especially at night, when they place barricades to stop unsuspecting travelers whom they rob and sometimes kill. "Rascal" falls far short of describing the nature of these criminals.

How lucky my driver, Stella, and I were that we were spared the trip down these very roads, my original destination, I thought.

Finally, our escorts in place, we returned unchallenged to the security of our ship where I found haven in my bunk, quickly in dreamland with an abundance of dream material to work with. That fourth day of September 1991 was a day to remember.

My hopes for a brilliant sunset were dampened the next day, when it clouded over early. My plan to cast the ashes at the southeasterly tip of New Ireland began to look doubtful. By the time we hoisted anchor, the wind had started to pick up. I prepared to stand by on deck with the little bag in my pocket an hour or so after dinner until we reached the place.

The wind increased steadily as we passed Duke of York Island and approached Cape Gazelle. The seas rose even more and by dinner time we were headed directly into the heart of a storm they said originated over Australia and would bear down with extreme force all the way to the Philippines. I found it difficult to stand. I gave up the idea of waiting on deck. I thought I would see Warangoi Bay as we passed. It had grown dark. Any thoughts of waiting on deck and watching landmarks pass by were out of the question. The storm raged and the ship barely made way as the huge waves picked up the bow and sent it crashing down. I could hardly stand or negotiate the passage to my stateroom. I sank into my bunk, tired and dizzy from any attempts to keep my balance, obvious symptoms of motion sickness.

As I lay there, thoughts went through my mind, *If this is a message telling me I've chosen the wrong place to drop ashes, I hear it loud and clear. I have never experienced such a storm at sea nor gotten a clearer message.*

Epilogue: Ashes

It took two days to reach the Treasuries, normally a one-day trip. Most of the passengers spent this time in their staterooms on their backs, like me. At 0800 we dropped anchor in a calmer sea in the protection of Mono Island. We were close to the lagoon where the Dumbo landed over fifty years ago and brought me to the safety and protection of USS *Coos Bay*.

After breakfast, we made our first landings on Zodiacs under a complete overcast, in sporadic rain showers. We walked along a path through the dripping jungle to the end of an old, unused airstrip, probably the one that carried the ambulance plane I boarded the day after my rescue. We saw the hulks of wrecked old warplanes along the way. On the windward side of the island we came upon an open sea that raged with violently roaring surf.

As our ship headed southeasterly down "The Slot" between islands that defined the Solomons in the War years, we continued to buck the storm, rough seas, heavy winds, dark overcast.

Our next stop, again a protected cove between little Islands, offered the opportunity to visit a small museum that bids homage to the memory of John F. Kennedy and to the tiny island which harbored him and his crew from PT109 after it had been rammed and sunk by a Japanese destroyer. The tiny island appealed to me. I had the small bag in my shirt pocket. *This might be the place.* Only five or six of us boarded the Zodiac. The Island, named after Kennedy, was unoccupied and so small it could be crossed in about two minutes. Trees and jungle plants offered privacy. I could scatter ashes discreetly in its center.

One of my shipmates, a woman, wanted to take my picture. I posed in my authentic adventure gear, shorts, short sleeved sport shirt topped with one of those French Army bush hats with the brim snapped up on one side.

That taken care of, I excused myself and headed for the bushes and the interior of the island. I reached for the bag and prepared for a little ritual. Something bothered me about this, though: it was so lonely there. She preferred to be with people, didn't cherish solitude as I did. All those other places I had suggested in the memorial service were people places: the Champs Elysee, the Louvre, Musee d'Orsay, Monet's Gardens... places where good people get together to do good things. *I just can't put her ashes in this lonely place.* I replaced the bag in my pocket and returned, thoughtfully, to the Zodiac.

And so it went, on down the island chains, finding significant places but not necessarily appropriate ones for this special event. On Guadalcanal we visited Henderson Field, an active strip now called Henderson Airport and Red Beach near the old Quonset hut named Mob Eight, the mobile hospital where I was treated which became a school after the war. Our native tour guide attended classes there as a boy.

At dawn we passed the eastern end of Savo island, where a number of sea battles were fought; we were over Iron Bottom Sound, where fighting vessels of both navies lie at its bottom. Preparations had been made for a ceremony to drop a memorial wreath for all the men who lost their lives there. Again, thoughts of ashes came to mind, but were hastily dismissed as I saw visions of dead men and wrecked ships—not a pretty sight.

Later that morning we finally got a brief sun break as we went ashore on Tulagi. In the afternoon we anchored nearby off a tiny idyllic looking island named Ghavutu where we enjoyed snorkeling on its colorful reefs. But the clouds returned, the winds rose and we moved into another storm late in the afternoon, as *World Discoverer* headed for Rennell Island.

As we sailed southward we approached some of the sources for James Michener's *Tales of the South Pacific:* the fictional Bali Hai, the Frenchman's plantation on Espiritu Santo in the Vanuatu group. This suggested romance and lovely places. I remembered the Blue Hole where we swam during the war. We called it Dorothy Lamour pool. It lay only a kilometer north of the runway at the base of a waterfall: an unusually deep turquoise, it lay like a jewel in the jungle near Turtle Bay Fighter Strip.

As we climbed aboard a tour bus at Luganville, I heard we would be going to one of the old airfields. The original itinerary had not even hinted at this. *Could The Turtle Bay strip be one of the old airfields?* I had a whole month of memories from there. *We could even visit the Blue Hole!*

About an hour later we were taking pictures at the end of the runway at Turtle Bay and a few minutes later some of the younger, more daring members were plunging into the Blue Hole with their clothes on. What a surprise that was! I got another surprise when I reached for my pocket, found it empty and remembered I changed shirts in the morning and left the little bag aboard. This is the closest I had come to finding just the right place, and I blew it.

Back aboard the *World Discoverer,* I began to realize why it has that name: one discovery after another seemed to unfold from the day of embarkation as we progressed southward. I still had the ashes but I believed the right place for them would present itself in a timely manner. We were to have three more days at sea and one at a resort on Lautoka in the Fiji group.

We visited more villages on Vanuatu islands. Besides our historian, we had an anthropologist aboard. He had actually lived with the natives in a number of the villages and knew many of their citizens by name. They entertained us with rituals, songs, and dances and laid out products of their many arts and crafts for our pleasure and purchase.

In a moderate sea, anchored off Tomman Island, we snorkeled around an offshore reef; in the rain, we experienced the sensation of warm water below, cold drops pelting our backs.

Two days later on the open sea we still had the lingering remnants of the storm; heavy seas and squalls prevailed most of this, our last full day aboard, but late afternoon brought a break in the clouds, producing just the right foil for a setting sun, a setting that delivered a display of prismatic colors, streaks of light shooting through dynamic cloud shapes.

The beauty of the sunset evoked reminders of Rosalind's ashes again. I looked around. *Is this the place?* The sky was breathtaking, glowing atop the deep blue Pacific but something said, *not yet.* The ocean looked vast and empty, too much open sea, lonely, with no landmarks. *I'll wait for a beautiful sunrise, with islands nearby and people having fun.*

At dawn I rose early in anticipation of a possible dynamic sunrise to match last night's dramatic display, but the sky was cloudless; we had left the storm behind. The sky was a brilliant blue, not multi-colored, as we approached the channel running between the Fiji Isles. The climate felt more arid, much like the desert atmosphere of Southern California... *her home.* We were approaching the place just right for her remains.

I found a secluded spot by the rail. According to a chart on the bridge, our destination, a resort on Lautoka, would be on our starboard side. Then I saw it: a beach with joyful sunbathers, in multicolored wear, boats and boards with colorful sails and brilliant parachutes aloft carrying daring ones behind speeding boats. Passengers embarking on an excursion vessel with striped awnings were about to move out into the channel to visit other islands in the group—a lot of happy people having fun, a good place to drop Roz off, a good place to think of her, and so I did. I loosened the drawstrings of the little bag and eased her over the side, into a tranquil sea.

The ship made a slight starboard turn as its bow wave slid out away from the side creating a marker that I could watch as long as it endured. The little wave seemed bent on immortality. I watched as it stood, a little white cap that ceremoniously sent back a secret salute. I never saw a bow wave last so long. It was gone only when it was too far away to see anymore. In my mind's eye, it stands forever.

F4U CORSAIR
Courtesy of the Stokes Collection, Carmel, California 1-800-359-4644